# The National Aeronautics and Space Administration

# KNOW YOUR GOVERNMENT

# The National Aeronautics and Space Administration

Tom D. Crouch

CHELSEA HOUSE PUBLISHERS

On the cover: On February 7, 1984, *Challenger* astronaut Bruce McCandless II tests a hand-controlled device, called the manned maneuvering unit, during his space walk. Previously, astronauts had to use restrictive tethers when moving around the exterior of the spacecraft.

Frontispiece: A bird's-eye view of the *Apollo 8/Saturn 5* space vehicle during its roll-out from the Vehicle Assembly Building at Kennedy Space Center in Florida in 1968.

**Chelsea House Publishers**

Editor-in-Chief: Nancy Toff
Executive Editor: Remmel T. Nunn
Managing Editor: Karyn Gullen Browne
Copy Chief: Juliann Barbato
Picture Editor: Adrian G. Allen
Art Director: Maria Epes
Manufacturing Manager: Gerald Levine

**Know Your Government**

Senior Editor: Kathy Kuhtz

**Staff for THE NATIONAL AERONAUTICS AND SPACE ADMINISTRATION**

Copy Editor: Brian Sookram
Deputy Copy Chief: Mark Rifkin
Editorial Assistant: Gregory R. Rodríguez
Picture Research: Dixon & Turner Associates, Inc.
Picture Coordinator: Melanie Sanford
Assistant Art Director: Loraine Machlin
Senior Designer: Noreen M. Lamb
Production Manager: Joseph Romano
Production Coordinator: Marie Claire Cebrián

3  5  7  9  8  6  4

**Library of Congress Cataloging-in-Publication Data**

Crouch, Tom D.
  The National Aeronautics and Space Administration / Tom D. Crouch.
    p.   cm.—(Know your government)
  Includes bibliographical references.
  Summary: Surveys the history of the National Aeronautics and Space Administration and describes its structure, current functions, and influence in our society.
  ISBN 1-55546-120-4
        0-7910-0901-7 (pbk.)
  1. United States.   National Aeronautics and Space Administration—Juvenile literature.   [1. United States.   National Aeronautics and Space Administration.]
  I. Title.   II. Series: Know your government (New York, N.Y.)          89-25389
TL521.C655 1990                                                                    CIP
353.0087'78—dc20                                                                   AC

# CONTENTS

# KNOW YOUR GOVERNMENT

CHELSEA HOUSE PUBLISHERS

# INTRODUCTION

# Government: Crises of Confidence

## Arthur M. Schlesinger, jr.

From the start, Americans have regarded their government with a mixture of reliance and mistrust. The men who founded the republic did not doubt the indispensability of government. "If men were angels," observed the 51st Federalist Paper, "no government would be necessary." But men are not angels. Because human beings are subject to wicked as well as to noble impulses, government was deemed essential to assure freedom and order.

At the same time, the American revolutionaries knew that government could also become a source of injury and oppression. The men who gathered in Philadelphia in 1787 to write the Constitution therefore had two purposes in mind. They wanted to establish a strong central authority and to limit that central authority's capacity to abuse its power.

To prevent the abuse of power, the Founding Fathers wrote two basic principles into the new Constitution. The principle of federalism divided power between the state governments and the central authority. The principle of the separation of powers subdivided the central authority itself into three branches—the executive, the legislative, and the judiciary—so that "each may be a check on the other." The *Know Your Government* series focuses on the major executive departments and agencies in these branches of the federal government.

The Constitution did not plan the executive branch in any detail. After vesting the executive power in the president, it assumed the existence of "executive departments" without specifying what these departments should be. Congress began defining their functions in 1789 by creating the Departments of State, Treasury, and War. The secretaries in charge of these departments made up President Washington's first cabinet. Congress also provided for a legal officer, and President Washington soon invited the attorney general, as he was called, to attend cabinet meetings. As need required, Congress created more executive departments.

Setting up the cabinet was only the first step in organizing the American state. With almost no guidance from the Constitution, President Washington, seconded by Alexander Hamilton, his brilliant secretary of the treasury, equipped the infant republic with a working administrative structure. The Federalists believed in both executive energy and executive accountability and set high standards for public appointments. The Jeffersonian opposition had less faith in strong government and preferred local government to the central authority. But when Jefferson himself became president in 1801, although he set out to change the direction of policy, he found no reason to alter the framework the Federalists had erected.

By 1801 there were about 3,000 federal civilian employees in a nation of a little more than 5 million people. Growth in territory and population steadily enlarged national responsibilities. Thirty years later, when Jackson was president, there were more than 11,000 government workers in a nation of 13 million. The federal establishment was increasing at a faster rate than the population.

Jackson's presidency brought significant changes in the federal service. He believed that the executive branch contained too many officials who saw their jobs as "species of property" and as "a means of promoting individual interest." Against the idea of a permanent service based on life tenure, Jackson argued for the periodic redistribution of federal offices, contending that this was the democratic way and that official duties could be made "so plain and simple that men of intelligence may readily qualify themselves for their performance." He called this policy rotation-in-office. His opponents called it the spoils system.

In fact, partisan legend exaggerated the extent of Jackson's removals. More than 80 percent of federal officeholders retained their jobs. Jackson discharged no larger a proportion of government workers than Jefferson had done a generation earlier. But the rise in these years of mass political parties gave federal patronage new importance as a means of building the party and of rewarding activists. Jackson's successors were less restrained in the distribu-

8

tion of spoils. As the federal establishment grew—to nearly 40,000 by 1861—the politicization of the public service excited increasing concern.

After the Civil War the spoils system became a major political issue. High-minded men condemned it as the root of all political evil. The spoilsmen, said the British commentator James Bryce, "have distorted and depraved the mechanism of politics." Patronage, by giving jobs to unqualified, incompetent, and dishonest persons, lowered the standards of public service and nourished corrupt political machines. Office-seekers pursued presidents and cabinet secretaries without mercy. "Patronage," said Ulysses S. Grant after his presidency, "is the bane of the presidential office." "Every time I appoint someone to office," said another political leader, "I make a hundred enemies and one ingrate." George William Curtis, the president of the National Civil Service Reform League, summed up the indictment. He said,

> The theory which perverts public trusts into party spoils, making public
> employment dependent upon personal favor and not on proved merit,
> necessarily ruins the self-respect of public employees, destroys the
> function of party in a republic, prostitutes elections into a desperate
> strife for personal profit, and degrades the national character by lower-
> ing the moral tone and standard of the country.

The object of civil service reform was to promote efficiency and honesty in the public service and to bring about the ethical regeneration of public life. Over bitter opposition from politicians, the reformers in 1883 passed the Pendleton Act, establishing a bipartisan Civil Service Commission, competitive examinations, and appointment on merit. The Pendleton Act also gave the president authority to extend by executive order the number of "classified" jobs—that is, jobs subject to the merit system. The act applied initially only to about 14,000 of the more than 100,000 federal positions. But by the end of the 19th century 40 percent of federal jobs had moved into the classified category.

Civil service reform was in part a response to the growing complexity of American life. As society grew more organized and problems more technical, official duties were no longer so plain and simple that any person of intelligence could perform them. In public service, as in other areas, the all-round man was yielding ground to the expert, the amateur to the professional. The excesses of the spoils system thus provoked the counter-ideal of scientific public admin-istration, separate from politics and, as far as possible, insulated against it.

The cult of the expert, however, had its own excesses. The idea that administration could be divorced from policy was an illusion. And in the realm of policy, the expert, however much segregated from partisan politics, can never attain perfect objectivity. He remains the prisoner of his own set of values. It is these values rather than technical expertise that determine fundamental judgments of public policy. To turn over such judgments to experts, moreover, would be to abandon democracy itself; for in a democracy final decisions must be made by the people and their elected representatives. "The business of the expert," the British political scientist Harold Laski rightly said, "is to be on tap and not on top."

Politics, however, were deeply ingrained in American folkways. This meant intermittent tension between the presidential government, elected every four years by the people, and the permanent government, which saw presidents come and go while it went on forever. Sometimes the permanent government knew better than its political masters; sometimes it opposed or sabotaged valuable new initiatives. In the end a strong president with effective cabinet secretaries could make the permanent government responsive to presidential purpose, but it was often an exasperating struggle.

The struggle within the executive branch was less important, however, than the growing impatience with bureaucracy in society as a whole. The 20th century saw a considerable expansion of the federal establishment. The Great Depression and the New Deal led the national government to take on a variety of new responsibilities. The New Deal extended the federal regulatory apparatus. By 1940, in a nation of 130 million people, the number of federal workers for the first time passed the 1 million mark. The Second World War brought federal civilian employment to 3.8 million in 1945. With peace, the federal establishment declined to around 2 million by 1950. Then growth resumed, reaching 2.8 million by the 1980s.

The New Deal years saw rising criticism of "big government" and "bureaucracy." Businessmen resented federal regulation. Conservatives worried about the impact of paternalistic government on individual self-reliance, on community responsibility, and on economic and personal freedom. The nation in effect renewed the old debate between Hamilton and Jefferson in the early republic, although with an ironic exchange of positions. For the Hamiltonian constituency, the "rich and well-born," once the advocate of affirmative government, now condemned government intervention, while the Jeffersonian constituency, the plain people, once the advocate of a weak central government and of states' rights, now favored government intervention.

In the 1980s, with the presidency of Ronald Reagan, the debate has burst out with unusual intensity. According to conservatives, government intervention abridges liberty, stifles enterprise, and is inefficient, wasteful, and arbitrary. It disturbs the harmony of the self-adjusting market and creates worse troubles than it solves. Get government off our backs, according to the popular cliché, and our problems will solve themselves. When government is necessary, let it be at the local level, close to the people. Above all, stop the inexorable growth of the federal government.

In fact, for all the talk about the "swollen" and "bloated" bureaucracy, the federal establishment has not been growing as inexorably as many Americans seem to believe. In 1949, it consisted of 2.1 million people. Thirty years later, while the country had grown by 70 million, the federal force had grown only by 750,000. Federal workers were a smaller percentage of the population in 1985 than they were in 1955—or in 1940. The federal establishment, in short, has not kept pace with population growth. Moreover, national defense and the postal service account for 60 percent of federal employment.

Why then the widespread idea about the remorseless growth of government? It is partly because in the 1960s the national government assumed new and intrusive functions: affirmative action in civil rights, environmental protection, safety and health in the workplace, community organization, legal aid to the poor. Although this enlargement of the federal regulatory role was accompanied by marked growth in the size of government on all levels, the expansion has taken place primarily in state and local government. Whereas the federal force increased by only 27 percent in the 30 years after 1950, the state and local government force increased by an astonishing 212 percent.

Despite the statistics, the conviction flourishes in some minds that the national government is a steadily growing behemoth swallowing up the liberties of the people. The foes of Washington prefer local government, feeling it is closer to the people and therefore allegedly more responsive to popular needs. Obviously there is a great deal to be said for settling local questions locally. But local government is characteristically the government of the locally powerful. Historically, the way the locally powerless have won their human and constitutional rights has often been through appeal to the national government. The national government has vindicated racial justice against local bigotry, defended the Bill of Rights against local vigilantism, and protected natural resources against local greed. It has civilized industry and secured the rights of labor organizations. Had the states' rights creed prevailed, there would perhaps still be slavery in the United States.

The national authority, far from diminishing the individual, has given most Americans more personal dignity and liberty than ever before. The individual freedoms destroyed by the increase in national authority have been in the main the freedom to deny black Americans their rights as citizens; the freedom to put small children to work in mills and immigrants in sweatshops; the freedom to pay starvation wages, require barbarous working hours, and permit squalid working conditions; the freedom to deceive in the sale of goods and securities; the freedom to pollute the environment—all freedoms that, one supposes, a civilized nation can readily do without.

"Statements are made," said President John F. Kennedy in 1963, "labelling the Federal Government an outsider, an intruder, an adversary. . . . The United States Government is not a stranger or not an enemy. It is the people of fifty states joining in a national effort. . . . Only a great national effort by a great people working together can explore the mysteries of space, harvest the products at the bottom of the ocean, and mobilize the human, natural, and material resources of our lands."

So an old debate continues. However, Americans are of two minds. When pollsters ask large, spacious questions—Do you think government has become too involved in your lives? Do you think government should stop regulating business?—a sizable majority opposes big government. But when asked specific questions about the practical work of government—Do you favor social security? unemployment compensation? Medicare? health and safety standards in factories? environmental protection? government guarantee of jobs for everyone seeking employment? price and wage controls when inflation threatens?—a sizable majority approves of intervention.

In general, Americans do not want less government. What they want is more efficient government. They want government to do a better job. For a time in the 1970s, with Vietnam and Watergate, Americans lost confidence in the national government. In 1964, more than three-quarters of those polled had thought the national government could be trusted to do right most of the time. By 1980 only one-quarter was prepared to offer such trust. But by 1984 trust in the federal government to manage national affairs had climbed back to 45 percent.

Bureaucracy is a term of abuse. But it is impossible to run any large organization, whether public or private, without a bureaucracy's division of labor and hierarchy of authority. And we live in a world of large organizations. Without bureaucracy modern society would collapse. The problem is not to abolish bureaucracy, but to make it flexible, efficient, and capable of innovation.

Two hundred years after the drafting of the Constitution, Americans still regard government with a mixture of reliance and mistrust—a good combination. Mistrust is the best way to keep government reliable. Informed criticism is the means of correcting governmental inefficiency, incompetence, and arbitrariness; that is, of best enabling government to play its essential role. For without government, we cannot attain the goals of the Founding Fathers. Without an understanding of government, we cannot have the informed criticism that makes government do the job right. It is the duty of every American citizen to know our government—which is what this series is all about.

On October 4, 1957, the Soviet Union launched the world's first artificial satellite, Sputnik 1, ushering in the beginning of the space age. A replica of the
satellite was displayed in the Moscow Exhibition Hall later that year.

# ONE

# The Space Age

The news burst like a thunderclap on the evening of October 4, 1957. It was just after six o'clock on the East Coast of the United States when the Soviet newspaper *Pravda* announced the successful launch of *Sputnik 1*, the world's first artificial earth satellite.

In Cambridge, Massachusetts, a small symphony orchestra, its members drawn from the staffs of the Harvard College Observatory and the Smithsonian Astrophysical Observatory, had just begun a rehearsal of the Russian master-piece *Peter and the Wolf*. One by one, key officials were called away from their music stands, briefed, and sent off to begin calculating the orbit of the spacecraft, setting a satellite tracking program in motion and responding to the calls flooding into the switchboard from the press and public.

Texas senator Lyndon B. Johnson was throwing one of his legendary barbecues at the LBJ ranch that evening. After dinner he strolled down toward the Pedernales River with a few of his guests and looked up at the sky. "In the Open West," he wrote later, "you learn to live closely with the sky. But now, somehow, in some new way, the sky seemed almost alien. I . . . remember the profound shock of realizing that it might be possible for another nation to achieve technological superiority over this great country of ours."

The scientists and engineers of the Army Ballistic Missile Agency in Huntsville, Alabama, were in the middle of a reception honoring Secretary of Defense Neil McElroy when the news of *Sputnik* arrived. Wernher von Braun,

15

the man who had perhaps done more than any other to set human beings on the path to space, was first stunned, then angry. "We knew they were going to do it," he told the secretary. "For God's sake, turn us loose and let us do something. We can put up a satellite in sixty days, Mr. McElroy! Just give us the green light and sixty days." Von Braun's superior, General John Medaris, stepped forward. "No, Wernher," he remarked, "ninety days."

The world would never be quite the same after October 4, 1957. *Sputnik 1*, a polished metal sphere no bigger than a beach ball, had changed things forever. Human beings, explorers by their very nature, had set off on their ultimate voyage of discovery.

Many Americans, however, found it difficult to share the sense of excitement that gripped the rest of the world during the first weeks and months of the space age. The United States and the Soviet Union had emerged from World War II as the most powerful nations on the globe, and the bitterest of rivals. The uneasy wartime alliance between the two superpowers gave way to mutual hostility and distrust after 1945. International tension became a permanent fact of life as the Soviets tightened their grip on the nations of Eastern Europe, developed atomic weapons to match those of the United States, and extended their influence into Asia, Africa, and Latin America.

The threat of a nuclear holocaust prevented a direct armed confrontation between the United States and the Soviet Union. Instead, the two nations struggled to win the support and friendship of developing nations through programs of economic and military assistance and demonstrations of national power and technical prowess. Given the atmosphere of cold war rivalry, any Soviet triumph was, by definition, an American defeat.

The launching of *Sputnik* was a Soviet masterstroke. Almost everyone had assumed that the United States would lead the way into space. For months newspapers had been filled with stories about American plans to launch a Vanguard earth satellite during the International Geophysical Year (IGY), 1957–58—a period during which scientists around the globe would concentrate on achieving a better understanding of the earth, its atmosphere, and near-earth space. The Soviets announced that they would also launch a satellite during the IGY, but they avoided publicity, cloaking their efforts in secrecy. As a result, *Sputnik* caught almost everyone by surprise.

For Americans, there was worse to come. On November 3, 1957, the 1,118-pound *Sputnik 2* carried the world's first space traveler, a dog named Laika, into orbit. American spirits hit rock bottom less than a month later, when the rocket that was to have launched *Vanguard TV-3*, America's first earth satellite, was destroyed in a catastrophic explosion on the pad.

*On December 6, 1957, the United States's first attempt to launch a satellite,* Vanguard TV-3, *into earth orbit ended on the launchpad with an explosion of the rocket's booster. Despite the failure of the launch, the United States was eager to take up the Soviet Union's challenge to a "space race."*

Shaken and bewildered by their inability to match the repeated demonstrations of Soviet prowess, Americans began to ask searching questions. Was *Sputnik* proof that the Soviets had forged ahead in the vital field of missile technology? Did it point to a basic flaw in the American system? Were Americans too soft, too materialistic? Had they lost their will to compete?

Elder statesman and business leader Bernard Baruch summed up the doubts and fears of his fellow citizens with a touch of dark humor. "If America ever crashes," he remarked, "it will be in a two-tone convertible." Senator Styles Bridges put it more bluntly: Americans would have to worry less about "the height of the tail fin in the new car and be more prepared to shed blood, sweat and tears if this country and the free world are to survive."

President Dwight D. Eisenhower, only mildly surprised by *Sputnik*, was stunned to discover that it seemed to matter so much to so many Americans. Administration officials did their best to put matters into perspective and restore confidence, but Congress and the press continued to demand action. The Soviets had challenged the United States to a "space race." With national power and prestige at stake and the whole world watching, Americans were eager to take up the challenge.

Apollo 11 *astronaut Edwin "Buzz" Aldrin poses for a photograph beside a U.S. flag. NASA's all-out effort in the 1960s to land a man on the moon before the end of the decade succeeded; the Lunar Module* Eagle *(left) touched down on the lunar surface on July 20, 1969.*

# TWO

# NASA Today

$A$ new federal agency, the National Aeronautics and Space Administration (NASA), grew out of the frustration and uncertainty of the Sputnik era. From the moment President Eisenhower signed the National Aeronautics and Space Act creating the new agency on July 29, 1958, great things were expected of the men and women of NASA. They were asked to achieve what most people had thought impossible only a few years before—to fly into space—and they were expected to do it with the entire world looking over their shoulders.

Scientific research and the pursuit of the practical benefits of spaceflight were part of the NASA program from the outset, but there was never any serious doubt as to the agency's primary goal. NASA was in business to overtake the Soviet space effort and restore confidence in the technological strength of the United States.

Americans were willing to pay a considerable price to achieve that goal, a fact that was reflected in NASA's phenomenal early growth. The agency's budget soared from $300.9 million in 1959 to a whopping $5.27 billion in 1967, the peak spending year in preparation for the Apollo moon landings. Staffing climbed from 9,235 at the outset to an all-time high of 35,860 in 1967.

NASA's all-out effort succeeded. The space agency landed 12 human beings on the lunar surface and returned them safely to earth between 1969 and 1972. As far as most Americans were concerned, the "race" with the Soviets was over—and with it the need for extraordinary expenditures on the space program. As appropriations in support of the war in Vietnam and new domestic programs climbed to unprecedented levels, NASA's budget and personnel figures slumped. The space agency's funding fell to $3.039 billion in 1974. Employment declined every year between 1967 and 1982.

The approval of the Space Shuttle program in the early 1970s marked the beginning of a new era in the history of NASA. Agency officials argued that the shuttle, a reusable "space truck," would drastically reduce the cost of flying human beings into orbit and serve as the cornerstone for America's long-term future in space. Following the low point of the post-Apollo years, NASA entered an era of slow, steady growth in 1976, as Congress began to provide the funds with which to design, build, and operate the new spacecraft. In 1987 the agency received a record federal appropriation of $10.796 billion and employed 22,500 men and women at its 13 major research and test facilities, launch centers, and laboratories. Roughly the same number of people were employed by private businesses, universities, and other NASA contractors.

Today's NASA is a large and complex organization, responsible for almost all aspects of the U.S. civilian space enterprise. Most people think of the space agency in terms of the courageous exploits of the astronauts and the excitement of new discoveries in space. They seldom pause to consider the all-important task of administration. The work of administrators and planners who spend most of their time seated at a desk may not seem terribly exciting, but wise management of the money, time, talent, and other scarce resources made available to NASA is the first and most important ingredient of a successful space program. Any attempt to understand the past, present, or future of the American space effort must begin with a look at the way in which NASA is managed.

The nerve center of NASA is to be found in the Office of the Administrator, located in the headquarters building on Independence Avenue in Washington, D.C. The administrator of NASA is appointed by the president with the approval of the Senate. Responsible for every aspect of NASA operations, he or she is assisted by a deputy administrator, an executive officer, several advisory committees, and, as of 1990, 16 individuals operating at the level of associate or assistant administrator. Some of these officials are responsible for such specific management functions as handling legal problems, preparing budgets and tracking funds, contracting for goods and services, public relations, and long-range planning. Other associate administrators head the all-important NASA program offices. The numbers, names, and precise functions of the program offices have varied over the course of NASA's history, reflecting changes in the direction and future plans of the agency. As

*In 1989, Administrator Richard Truly (left) and Heinz Riesenhuber, federal minister for research and technology of the Federal Republic of Germany, sign a memorandum of understanding in Washington, D.C., enabling the launch of German payloads on the space shuttle. Truly, who flew on space shuttle missions in 1981 and 1983, became the eighth NASA administrator.*

of 1990, there were five such offices: Space Station; Space Operations; Aeronautics and Space Technology; Space Science and Applications; and Space Flight.

The associate administrator for Space Station heads the newest program office, with responsibility for planning and overseeing work on the manned space station that NASA hopes to construct in earth orbit during the 1990s. The agency has already signed agreements with several foreign partners willing to share the cost of developing this permanently occupied orbital facility.

Maintenance of NASA's worldwide space tracking and communications network belongs to the associate administrator for Space Operations. Composed of both ground stations and special communications and data relay satellites, the network is operated by men and women who are responsible for all communications with U.S. spacecraft, manned and unmanned. In addition, the office sometimes assists foreign nations with space communications. The associate administrator for Aeronautics and Space Technology operates a broad range of research and development programs designed to keep the

*At Langley Research Center in Hampton, Virginia, a section of an aircraft wing— mounted on a tinker- toy-like carriage—is propelled to 155 MPH to see what effect heavy rain has on wing performance.*

United States at the forefront of aerospace technology. The effort includes basic and theoretical research; practical efforts to solve current problems faced by the designers of airplanes and spacecraft; and the development and testing of systems for the future.

NASA conducts most of its aerospace technology research at three laboratory facilities—the Langley, Ames/Dryden, and Lewis research centers. The director of each center reports to the associate administrator for Aeronautics and Space Technology. Located in Hampton, Virginia, the Langley Research Center is NASA's oldest laboratory. Since 1917, when the National Advisory Committee for Aeronautics established the research center, the scientists and engineers of Langley have devoted themselves to the development of the world's most advanced aircraft systems. No longer content with flying airplanes higher, faster, and farther than anyone else, they are now searching for new ways to fly more safely, more efficiently, and with less impact on the environment. One of the most important projects now under way at Langley involves the development of a National Aero-Space Plane, a revolutionary winged transport craft that will fly at unprecedented speeds and altitudes. In addition to its other responsibilities, Langley also administers the Wallops Flight Facility on Wallops Island, Virginia, which conducts scientific research with instrumented rockets and enormous helium balloons.

The Ames Research Center in Mountain View, California, pursues the basic scientific research that will result in new aerospace technologies. Computer science and applications, flight simulation, aerodynamics, and astronomy are a few of the many special fields in which work is currently under way at Ames.

The Ames Research Center also manages NASA's Dryden Flight Test Center in Lancaster, California. Since the end of World War II the most advanced aircraft in the world have been put through their paces in the skies over Dryden and nearby Edwards Air Force Base. With its long runways set

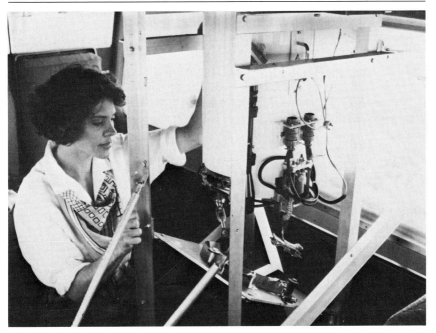

*A scientist at the Jet Propulsion Laboratory (JPL) in Pasadena, California, inspects an infrared radiometer, a highly sensitive instrument that measures thermal radiation at the ocean's surface. The infrared radiometer helps JPL scientists assess the reliability of radiation measurements received from weather satellites that are equipped with less sophisticated radiometers.*

# Robert Hutchings Goddard: America's Space Pioneer

The space age began as a dream in the minds of three men. A Russian, Konstantin Eduardovich Tsiolkovsky (1857–1935), laid down the first principles of space travel in *The Exploration of Cosmic Space with Reactive Devices*, published in 1911. Hermann Oberth (b. 1894) wrote a number of books on the subject between 1923 and 1929 and attracted a handful of brilliant young Germans to the field. But it was Robert Hutchings Goddard (1882–1945), the American pioneer, who combined theory with practice by designing and constructing the world's first liquid propellant rockets.

October 19, 1899, was a day that 16-year-old Robert Goddard would never forget. He was perched high up in a cherry tree that afternoon, trimming the branches and thinking about an exciting story he had just read—a tale of interplanetary travel that sent a string of questions racing through his mind. What would an interplanetary spacecraft look like? How would it function? He imagined such a craft descending from the sky and landing in the meadow at his feet. "I was a different boy when I descended that tree," he remarked many years later. "Existence at last seemed very purposive."

A brilliant student, Goddard was valedictorian of his high school class, graduated from the Worcester Polytechnic with honors, earned a doctorate in physics, then took a teaching job at Clark University, in Worcester, Massachusetts. But the budding college professor did not forget the dream of spaceflight. The questions that had intrigued him that afternoon in the cherry tree remained unanswered. How fast would a vehicle have to travel in order to leave the earth and journey into space? How could such a speed be attained? What sort of power plant could operate outside of the atmosphere?

Goddard provided the answers to those questions in a booklet entitled *A Method of Reaching Extreme Altitudes*, published by the Smithsonian Institution in 1919. But he was determined to also provide a practical demonstration of his theoretical principles. During World War I he experimented with a solid propellant rocket weapon for the army. He recognized, however, that liquid propellants would produce smoother combustion and a much higher exhaust velocity than available solids. Liquid hydrogen and liquid oxygen were the perfect fuel and oxidizer combination. Goddard was able to substitute gasoline for liquid hydrogen, but there was no substitute for liquid oxygen, and it had to be stored and used at a temperature of –298 degrees Fahrenheit. In addition, Goddard had to devise a way to move the propellant and oxidizer from their tanks to the combustion chamber where they could be mixed and ignited.

Overcoming these obstacles, Goddard launched the world's first liquid propellant rocket from his

aunt Effie's cabbage patch in Auburn, Massachusetts, on March 16, 1926. The rocket remained in the air for only 2½ seconds, achieving an altitude of 41 feet and landing 184 feet from the takeoff point. Still, it had flown higher and traveled farther than had the Wright brothers' airplane on its first flight.

With the assistance of Charles Abbot, secretary of the Smithsonian Institution, Goddard was able to continue his experiments at Fort Devons, Massachusetts, but the situation was far from ideal. A telephone call from Charles Lindbergh on a November afternoon in 1929 changed all of that. Catapulted to fame when he soloed the Atlantic in the *Spirit of St. Louis* two years before, Lindbergh had read about Goddard's experiments, and his problems, in the *New York Times*.

Fascinated, he arranged to meet Goddard and then convinced a wealthy acquaintance, Daniel Guggenheim, that the rocket work was worthy of support.

With Guggenheim funding and some additional grants Goddard established a laboratory, machine shop, and rocket launch facility near Roswell, New Mexico. Here he would spend most of the next 10 years, building and flying a series of larger and more complex liquid propellant rockets.

Progress was slow as Goddard struggled to perfect the pumps, guidance system, combustion chambers, and other elements that make up a rocket. By 1940 he had sent a great many rockets aloft, but none of them had reached an altitude of even one mile.

At the beginning of World War II, Goddard returned to the East Coast and continued to work on rocket programs until his death on August 10, 1945. NASA commemorated his achievements by naming its major satellite facility the Goddard Space Flight Center in his honor.

"Sitting in his home in Worcester, Massachusetts, in 1929," Charles Lindbergh recalled many years later, "I listened to Robert Goddard outline his ideas for the future development of rockets—what might be practically expected, what might be eventually achieved. Thirty years later, watching a giant rocket rise above the air force test base at Cape Canaveral, I wondered whether he was dreaming then, or I was dreaming now."

on the enormous sediment flats of Muroc Dry Lake, Dryden/Edwards is the primary landing site for the space shuttle.

The Lewis Research Center (LRC), located in Cleveland, Ohio, is named for George Lewis, a pioneering leader of the National Advisory Committee for Aeronautics, NASA's predecessor organization. The LRC takes the lead in the development of aircraft and space propulsion systems, space power, and satellite communications systems. At present, scientists and engineers at Lewis are concentrating on creating the equipment that will generate the electrical power required to operate the space station now taking shape on NASA drawing boards.

NASA's unmanned space effort is the province of the associate administrator for Space Science and Applications. Most of the work of this office is divided between two organizations: the Goddard Space Flight Center and the Jet Propulsion Laboratory (JPL). The scientists and engineers of the Goddard Space Flight Center in Greenbelt, Maryland, design, build, and operate the earth-orbiting scientific and applications satellites that have opened the frontiers of new knowledge and put the unique vantage point of space to work for mankind.

The men and women of the JPL in Pasadena, California, concentrate on deep space missions and spacecraft. (Deep space is the region beyond Earth's atmosphere, including space outside the Solar System.) A government-owned facility staffed by employees of the California Institute of Technology, the JPL also operates the Deep Space Communications Complex, a major station in the Deep Space Network that maintains constant contact with unmanned space probes on missions to other planets.

The associate administrator for Space Flight oversees the most visible element of the NASA program—manned space operations—and supervises the best-known NASA centers. The Marshall Space Flight Center (MSFC) in Huntsville, Alabama, manages the development of American space launch vehicles. Founded by the U.S. Army in 1950 as the Ordnance Guided Missile Center at the Redstone Arsenal, the facility was the longtime headquarters of Wernher von Braun and the team of engineers who sent the first American satellite into orbit in 1958. NASA took over the center in 1960 and renamed the facility in honor of George C. Marshall, who served as U.S. Army chief of staff during World War II and as secretary of state and later secretary of defense under President Harry Truman. The MSFC also supervises operations at the Michaud Assembly Facility, near New Orleans, Louisiana, where the external propellant tanks for the space shuttle are assembled and tested.

*Pad 39A, at Kennedy Space Center, Cape Canaveral, Florida, is the launch site for the space shuttle. Kennedy Space Center, which began operations in 1951 as the experimental missile firing branch of the Ordnance Guided Missile Center, became a NASA facility in 1960.*

27

The rockets developed by the experts at MSFC are sent aloft from the Kennedy Space Center, Cape Canaveral, Florida (renamed Cape Kennedy during the years 1963–73). Like the MSFC, the Kennedy Space Center began as a U.S. Army operation, a "Missile Firing Laboratory" for testing the rockets designed at Huntsville. NASA took charge of the facility in 1960 and began a program of expansion that transformed this isolated coastal area into "Spaceport, USA."

NASA established the Manned Space Center in Houston, Texas, in 1961. Renamed the Johnson Space Center (JSC) in honor of President Lyndon B. Johnson in 1973, the center is responsible for the design of manned spacecraft and serves as home base for astronauts, mission specialists, and other crew members. The JSC had primary responsibility for the overall development of the space shuttle and is now heavily involved in planning for a future space station.

For many years, NASA conducted tests of its rocket propulsion systems at yet another Office of Space Flight facility, the National Space Technology Laboratories in Bay St. Louis, Mississippi. Now renamed the John C. Stennis Space Center, in honor of one of NASA's important congressional supporters, the facility remains the agency's primary rocket engine test site. In addition, personnel of the Stennis center are actively engaged in scientific research that will lead to an improved understanding of the earth and its oceans.

NASA remains one of the best-known and most visible agencies of the federal government. It has provided Americans of the late 20th century with some of their proudest memories—space-suited astronauts saluting the Stars and Stripes on the dusty surface of the moon; a glimpse of the pale red landscape of Mars as seen by the Viking Lander spacecraft; stunning images of Saturn's rings and the multicolored clouds of Jupiter returned by a pair of Voyager spacecraft on the first leg of a journey that would carry them far beyond the limits of the Solar System.

But if NASA has tasted moments of heady triumph, it has also suffered through periods of crisis—the loss of the first Apollo crew in a catastrophic fire on the launch pad; a disastrous explosion in space that threatened the life of the *Apollo 13* astronauts; and the death of seven crew members aboard the space shuttle *Challenger*.

NASA weathered those storms, but it continues to face serious questioning from Congress, the press, and the public. Is the space program worth its enormous price tag? Did the effort required to place human beings on the moon prevent the formation of a more balanced program of space exploration and

*A section of* Challenger's *right wing is unloaded at Kennedy Space Center following the space shuttle explosion 73 seconds after takeoff on January 28, 1986. After the tragedy in which all seven crew members were killed, NASA grounded its shuttles and worked to correct design, safety, and management problems.*

development? Did the loss of the *Challenger* indicate deep-seated flaws in the management of NASA programs?

There are no easy answers to most of those questions. After three decades of American activity in space, however, one thing is clear. In good times and bad, the nation's space agency has never been far from the center of public attention. The hopes and dreams of the nation still ride on the outcome of every major launch.

*Charles Doolittle Walcott, head of the Smithsonian Institution from 1907 to 1927, helped convince Congress of the need for a program of government-sponsored research in the field of aerodynamics. In 1915, President Woodrow Wilson signed a bill creating the National Advisory Committee for Aeronautics (NACA).*

# THREE

# NACA: The Old Order

$T$he story of NASA begins in 1915, long before the launch of *Sputnik 1*. Like many of his fellow Americans, Charles Doolittle Walcott, head of the Smithsonian Institution, was worried about the state of the nation's defenses. The Great War in Europe, now almost a year old, would transform the countryside of France and Belgium into a ghastly proving ground for a new generation of weapons—first machine guns and long-range artillery, and later, toward the end of the war, poison gas and tanks. Submarines prowled coastal waters in search of prey, while giant zeppelin airships launched an aerial bombing campaign against England. Science and technology had given birth to a new and horrifying concept—"total" war.

Neutral America lagged far behind the warring powers of Europe in every branch of military technology, but Charles Walcott was especially concerned about the dismal condition of aviation in the United States. Like the machine gun and the submarine, the airplane was a product of American ingenuity. Two young bicycle makers from Ohio, Wilbur and Orville Wright, had made their first flight from a secluded North Carolina beach on December 17, 1903. With war clouds looming on the horizon, however, European leaders were quicker than their American counterparts to recognize the military potential of the new technology and to encourage its development. Officials in England, France, Germany, Italy, and Russia sponsored speed, altitude, and distance competitions, established aerial units within their armed forces, and created national

*On December 17, 1903, at Kill Devil Hill near Kitty Hawk, North Carolina, Orville Wright became the first person to successfully fly a self-propelled heavier-than-air machine. While his brother, Wilbur, watched, Orville flew 120 feet and ushered in the age of aviation.*

aeronautical laboratories to conduct programs of research and development that would unlock the secrets of flight. In the process, they forged the most revolutionary of all their new weapons.

By 1915, America scarcely qualified as a third-rate aeronautical power, and the gap was growing wider. Spurred on by the pressure of war, European engineers were building airplanes that flew higher, faster, farther, and with greater maneuverability. American designers, cut off from European research and combat experience by a curtain of wartime secrecy, had nowhere to turn for answers to their most basic questions. How were they to develop more effective wing shapes? What steps could be taken to reduce the air resistance, or drag, of their machines? Was it possible to design a better propeller? Could anything be done to solve the riddles of aircraft stability and control?

Convinced that any attempt to return the United States to a position of leadership in the air would have to begin by answering those questions, Charles Walcott set out in 1913 to win congressional approval for a program of government-sponsored research into the field of aerodynamics—the study of the complex interrelationship between the air and bodies of various shapes, sizes, and characteristics moving through it. Success came on March 3, 1915, when President Woodrow Wilson signed a bill creating a National Advisory Committee for Aeronautics (NACA).

Congress gave the NACA $5,000 for its first year of operations and charged the organization "to supervise and direct the scientific study of the problems of flight, with a view to their practical solution." It was a new kind of federal agency, designed to conduct basic research that would advance a technology regarded as essential to the defense of the nation. A committee of 12 leading aviation experts, appointed by the president to serve without pay, headed the organization. The committee made the basic decisions, providing leadership and direction for research. In addition, the members of the main committee would appoint special subcommittees to concentrate on specific technical issues. From top to bottom, it was an organization designed to stand above politics, making wise technical decisions and conducting research that would strengthen the American aviation industry, and thus the nation.

*During World War I, First Lieutenant Eddie Rickenbacker, America's "Ace of Aces" (an ace is an aviator who has shot down at least five enemy aircraft), is photographed with his plane in Moselle, France. Rickenbacker was one of the many distinguished men in aeronautics who served on the NACA.*

Some of the most distinguished men in American aeronautics served on the committee during its 43-year history. Orville Wright was a member for 28 years, longer than any other individual. Among the other longtime leaders of the NACA were Charles Lindbergh, the first pilot to solo the Atlantic; Henry H. "Hap" Arnold, commander of the U.S. Army Air Forces in World War II; Captain Eddie Rickenbacker, America's First World War "Ace of Aces"; James H. Doolittle, the best-known American speed flyer of his generation and leader of the famous Doolittle raid against Tokyo in 1942; and Karl T. Compton, one of the nation's leading physicists and educators.

The NACA came too late to have an impact on World War I. The United States finally entered the conflict in 1917 with an air arm consisting primarily of unarmed Curtiss JN-4D "Jenny" training aircraft. American pilots flew into combat aboard airplanes designed, and for the most part built, in Europe. The experience simply underscored the need for an organization like the NACA.

The committee's first order of business was to establish a laboratory, complete with wind tunnels, machine shops, specialized testing equipment, and an experimental flying field. That facility, the Langley Memorial Aeronautical Laboratory, opened its doors in Hampton, Virginia, on June 11, 1920. Named after aviation pioneer Samuel Pierpont Langley, Walcott's predecessor as head

*In 1934, 8 of the 12 members of the NACA attend an aircraft engineering research conference at Langley Field, Virginia. Seated (left to right) are Charles A. Lindbergh, Arthur B. Cook, Charles G. Abbot, Joseph S. Ames, Orville Wright, Edward P. Warner, Ernest J. King, and Eugene L. Vidal. Standing are: George Lewis, Henry J. E. Reid, and John F. Victory.*

*In 1931, a Vought 03U-1 airplane is tested in a wind tunnel at Langley Memorial Aeronautical Laboratory in Hampton, Virginia. The NACA laboratory, which also included machine shops, testing equipment, and an experimental flying field, opened in 1920.*

of the Smithsonian, the laboratory began with only three buildings. Over the next three decades it would grow to become one of the world's great research centers.

Led by committee members Joseph S. Ames of the Johns Hopkins University and Jerome C. Hunsaker, founder of the aeronautical engineering program at the Massachusetts Institute of Technology, the NACA focused its efforts on the study of aerodynamics, just as Charles Walcott had suggested. The committee imported talent and ideas from Europe and developed sophisticated new engineering tools, such as the variable-density wind tunnel, that opened new avenues to an understanding of the behavior of airplanes in flight.

By the end of its first 15 years, the NACA had demonstrated the practical value of basic research. The technical reports published by Langley engineers outlined the characteristics of various airfoils, or wing cross-sectional shapes; urged streamlining through the use of retractable landing gear, proper engine

35

placement, and fillets that blended wing and tail surfaces smoothly into the fuselage, or body, of the aircraft; explained the advantages of wing flaps and other high-lift devices; and explored the potential of new materials, such as the lightweight aluminum alloy, duralumin. The 1929 award of the prestigious Collier Trophy to a Langley team for the development of the NACA cowling, a streamlined engine housing that improved power plant cooling while reducing drag and increasing aircraft speed, suitably capped the first decade and a half of the agency's history.

American engineers listened to the NACA and learned. By the 1930s, a new generation of low-wing, streamlined, all-metal airplanes was flowing off their drawing boards. Aircraft such as the Douglas DC-3, the Boeing 247D, and the Sikorsky, Martin, and Boeing flying boats marked the return of the United States to a position of leadership in world aviation. NACA research had paid off. As one English engineering journal put it: "The present-day American position in all branches of aeronautical knowledge can, without doubt, be attributed to this far-seeing [NACA] policy. . . . "

The NACA experienced its "golden age" during the years between the wars. The $5,000 budget of 1915 climbed to just over $4 million by 1939. The staff grew from 181 employees in 1930 to 523 in 1939. There were new laboratory facilities as well. On September 14, 1939, NACA officials broke ground for what was to become the Ames Aeronautical Laboratory, at Moffett Field, a U.S. Navy air station near Mountain View, California.

Congress approved plans for a third NACA laboratory in June 1940. Located adjacent to the Cleveland Municipal Airport in Ohio, the new facility was fully equipped to study the three basic elements of aeronautical propulsion: engines, fuels, and lubricants. A center of American aircraft engine research during World War II, the lab was named in honor of George Lewis, a longtime committee member and one of the architects of the successful NACA research program, following his death in 1948.

Thanks in no small measure to the efforts of NACA scientists and engineers, the United States entered the 1940s far better prepared to fight an air war than it had been in 1917. "The Navy's famous fighters—the Corsair, Wildcat, and Hellcat—are possible only because they were based on fundamentals developed by the NACA," Secretary of the Navy Frank Knox remarked in 1943. "All of them use NACA wing sections, NACA cooling methods, NACA high lift devices. The great sea victories that have broken Japan's grip in the Pacific would not have been possible without the contributions of the NACA." Charles Walcott, who had died in 1927, would have been proud.

Top: *A Douglas DC-3, the first modern airliner, went into service in 1935. Designed to carry 21 passengers, it featured such advances as an all-metal stressed-skin construction, a radial engine, and retractable landing gear. Bottom: A Pan American World Airways flying boat, capable of carrying 48 passengers, began service across the Pacific in 1935. By the 1930s, NACA policy helped to boost the United States back to a position of leadership in world aviation.*

The NACA helped to develop and test American airplanes that flew into battle on every front. For the moment, basic research took a backseat to the solution of specific problems encountered in combat. Staff members struggled to understand why the controls of a fighter locked in a high-speed dive; why certain tail designs led to a loss of control; how the range, speed, and altitude of various aircraft types could be improved.

During the years after World War II, the men and women of the NACA turned their eyes to the future once again, considering the problems encountered by aircraft flying at extreme speeds and altitudes. When traditional wind tunnels failed to answer all of the questions posed by designers and pilots who dreamed of flying faster than sound, John Stack and his Langley colleagues developed a revolutionary new tunnel designed to study the behavior of aircraft operating at the speed of sound.

For other NACA personnel, the sky itself became a laboratory. Members of the Pilotless Aircraft Research Division pursued their studies at Wallops Island, an abandoned U.S. Navy test facility not far from the Langley Research Center. They dropped test shapes from high-flying aircraft and fired them into the atmosphere aboard rockets, recovering vital information via telemetry—radio messages broadcast from sensors on the models.

Out at Muroc Dry Lake, in the high desert of California, the military services and the NACA established an experimental flying field to test the most advanced aircraft in the world. It was here, on October 14, 1947, that Captain Chuck Yeager, aboard the Bell X-1, became the first man to fly faster than the speed of sound. Over the next decade, navy, air force, and NACA researchers pioneered new technologies and probed the limits of flight with the X series of experimental aircraft.

As the old barriers fell, new ones appeared in their place. Supersonic flight opened the door to the puzzle of transonic (beyond the speed of sound) drag. Engineers had assumed that a supersonic bullet provided the ideal shape for a high-speed aircraft. By 1951, however, wind tunnel tests of such shapes indicated that unexpected shock waves forming behind the wings increased drag. Richard Whitcomb of NACA Langley solved the problem.

Reasoning that a bullet does not have wings, a tail, or other protuberances, Whitcomb suggested that the fuselage should be pinched into a "coke bottle" shape to compensate for the cross-sectional area of those features. It worked. In the age of supersonic flight and billion-dollar government programs, progress was still dependent on a brilliant individual with a good idea.

Spaceflight did not come as a surprise to the NACA, but most staff engineers and scientists thought that flight beyond the atmosphere would be achieved

*Captain Chuck Yeager makes the world's first supersonic flight in the Bell X-1 on October 14, 1947. The NACA and the military services established an experimental flying field, now the Dryden Flight Test Center, at Muroc Dry Lake in California, where such advanced aircraft could be tested.*

gradually. It would be a step-by-step process, accomplished with future generations of exotic, rocket-propelled research aircraft capable of reaching ever-higher altitudes.

*Sputnik 1* changed all of that. President Eisenhower, beset by a storm of criticism for having allowed the Soviets to beat the United States into orbit, approved the development of a highly visible and competitive civilian space program. It would remain separate and distinct from ongoing military efforts, emphasizing the peaceful uses of space, performing scientific research, and exploring the new frontier.

Three federal agencies—the Department of Defense, the Atomic Energy Commission, and the National Advisory Committee for Aeronautics—bid for

# Milestones in Aeronautics

The first *A* in NASA has always stood for aeronautics. Building on a tradition of excellence in flight research established by the NACA, the men and women of the Langley, Lewis, and Ames Research Centers, and the Dryden Test Flight Center continue to search for new techniques to fly higher, faster, and farther with greater safety and economy.

Since the end of World War II, the knowledge flowing from NACA/NASA wind tunnels and laboratories has been tested in a variety of exotic research aircraft. Some, like the X-1 (the *X* stands for "experimental"), the first airplane to fly faster than the speed of sound, have explored the ultimate frontiers of flight. Others have tested specific design ideas, new materials, and advanced electronic systems.

Thirty years after its first flight, the North American X-15 remains the most successful and spectacular of the postwar research aircraft. NASA developed the X-15 in cooperation with the U.S. Air Force and test-flew three of these machines from March 25, 1960, to October 24, 1968. The sleek, rocket-propelled craft was the first true aerospace vehicle, reaching altitudes in excess of 67 miles and traveling 6.7 times the speed of sound. Neil Armstrong flew the X-15 for NASA before moving on to the Gemini and Apollo programs. Colonel Joe Engle was already wearing his astronaut wings when he made his first flight as a shuttle pilot. He had earned them aboard the X-15.

NASA continues to employ research aircraft as important tools for unlocking the secrets of winged flight. Some of the most interesting experimental airplanes flown in recent years include the following:

- The AD-1 (Ames-Dryden 1), a radio-controlled drone tested between 1979 and 1981. This small machine took off and flew at low speeds with its wing in a normal position at right angles to the fuselage. The wing could be "scissored," however, for high-speed runs, with the right tip pivoting forward and the left tip sweeping back to an angle of 60 degrees. With the wing in this position, the aircraft achieved greater speeds with reduced drag and lower fuel consumption.

- HiMAT, the Highly Maneuverable Aircraft Technology vehicle, another joint NASA/Air Force project. HiMAT is a remote-controlled aircraft roughly 44 percent the size of a full-scale fighter; it was designed to test new materials, techniques, and equipment to be employed on the next generation of high-performance fighters.

- The X-29A, a technology demonstrator with a forward swept wing, a forward elevator, and a computerized electronic control system.

*In 1960, the X-15 rocket airplane, designed to fly at speeds near 4,000 miles per hour and to altitudes above 50 miles, undergoes flight tests at Edwards Air Force Base in California.*

First test-flown in 1984, the X-29A is constructed with new materials better able to withstand the stresses and strains imposed by high-speed flight maneuvers.

- The M2, HL-10, and X-24, which derive part of their lift from the flow of air over the fuselage, as well as the wings. The information resulting from test flights with these "lifting bodies" played an important role in the design of the space shuttle.

- The XV-15 Tilt Rotor Research Aircraft, which pioneered new approaches to vertical flight and helped to pave the way for the development of V/STOL (Very Short Take-Off and Landing) vehicles.

Still other NASA programs are aimed at improving aircraft safety and efficiency. Whereas researchers at Langley study various aspects of flight instrumentation, safety, aircraft structures, and materials, the scientists and engineers of the Lewis Center focus their attention on reducing engine noise and pollution while improving fuel economy.

Seventy-five years ago the NACA set out to help the United States become a world aeronautical power. Today, NASA is still in the business of helping the nation retain leadership in the air.

41

Hugh Latimer Dryden, director of the NACA from 1947 to 1958, helped coordinate the aeronautical research (both on military and commercial aircraft) that was scattered around the country. In 1958, Dryden became NASA's first deputy administrator.

control of the new space program. James R. Killian, the president's special assistant for science and technology, and Senator Lyndon B. Johnson, chairman of the Preparedness Investigation Subcommittee of the Senate Armed Services Committee, were especially interested in the NACA, a civilian organization boasting an outstanding record of cooperation with both industry and the military and a history of success in flight research and development. On July 29, 1958, President Eisenhower signed Public Law 85-568, the National Aeronautics and Space Act, creating a new space agency on the foundation of the NACA.

The National Aeronautics and Space Administration (NASA) would have a new organizational structure and pursue very different goals from those of the NACA. The committee system that had served the agency so well was replaced with a streamlined bureaucracy. In August 1958, Eisenhower appointed the first NASA administrator—T. Keith Glennan, president of the Case Institute of Technology and a former Atomic Energy commissioner. Hugh Latimer Dryden, the last director of the NACA, would stay on as deputy administrator of the new agency.

Unlike the NACA, a basic research and problem-solving outfit, NASA would be responsible for every aspect of the nation's civilian space program. It would contract with the aerospace industry for launch vehicles, satellites, and manned spacecraft; construct launch and tracking facilities; conduct the research required to support those programs; and hire the individuals who would launch, monitor, and fly space missions.

On October 1, 1958, Glennan and Dryden gathered the headquarters staff together in the courtyard of their building, the Dolly Madison House, on Lafayette Square, near the White House, to celebrate the end of one era and the beginning of another. The venerable NACA, the world's most successful aeronautical research and development agency, was no more. In its place stood a new and untried organization. Ready or not, the National Aeronautics and Space Administration was in business.

NASA's first astronaut team, which was selected in 1959 for Project Mercury, included: (front, left to right) Walter M. "Wally" Schirra, Jr., Donald K. "Deke" Slayton, John H. Glenn, Jr., and Malcolm Scott Carpenter; (back, left to right) Alan B. Shepard, Jr., Virgil I. "Gus" Grissom, and L. Gordon Cooper.

# FOUR

# This New Ocean

$E$ight thousand men and women went to work for a new federal agency on the morning of October 1, 1958. The National Aeronautics and Space Administration would operate 3 laboratories (Langley, Ames, and Lewis) and 2 research stations (Wallops Island and Muroc) with an annual budget of $100 million. In addition, President Eisenhower had transferred the navy's Project Vanguard satellite program and selected army and air force lunar probe and rocket engine development efforts to NASA, along with an additional $100 million to support them. In order to avoid holding things up, NASA administrator T. Keith Glennan asked the Department of Defense to retain temporary control of these latter projects while he was organizing his own agency.

Things were already looking up for the United States. On January 31, 1958, Wernher von Braun and his colleagues sent the first U.S. satellite, *Explorer 1*, into orbit from Cape Canaveral, Florida, aboard a Jupiter-C rocket, a modified version of the army's Redstone ballistic missile. The navy team finally launched *Vanguard 1* on March 17, 1958.

The Soviets had led the way into space, but the first great scientific triumph of the era went to the Americans. *Explorer 1* made a dramatic discovery—rings of charged radiation particles trapped in Earth's magnetic field. *Vanguard 1* confirmed the existence of the Van Allen radiation belts, named in honor of James Van Allen, a University of Iowa scientist who played a key role in their discovery.

(*continued on page 50*)

45

# Wernher von Braun

Robert Goddard, Konstantin Tsiolkovsky, and Hermann Oberth had dreamed of space flight. More than any other single individual, Wernher von Braun made those dreams come true. His career spanned almost half a century, from the earliest experiments with liquid propellant rockets to the development of the giant Saturn 5.

He was born in Wirsitz, Germany, on March 23, 1912, the son of Baron Magnus von Braun, who served as an official in the Weimar government that ruled Germany following World War I. Baroness von Braun, an amateur astronomer, sparked her son's early interest in science. "For my confirmation I didn't get a watch or my first pair of long pants," von Braun remarked many years later. "I got a telescope. My mother thought it would make the best gift."

Leafing through an astronomy magazine some months later, he read an article describing an imaginary trip to the moon. "It filled me with a romantic urge," he recalled. "Interplanetary travel! Here was a task worth dedicating one's life to! Not just to stare through a telescope at the moon and planets but to soar through the heavens and actually explore the mysterious universe! I knew how Columbus had felt."

Young von Braun worked hard to acquire the sort of education that would be required of a latter-day Columbus, capping degrees in engineering and science with a doctorate in physics from the University of Berlin in 1934. His informal education in rocketry began when he joined the Verein für Ramschiffahrt (VfR—Society for Space Travel) while still in his teens. Founded by Hermann Oberth, the author of a pioneering text on spaceflight, the VfR was the most important of several small groups of rocket experimenters that had sprung up in Europe and America. By the fall of 1930 the sharp staccato roar of small liquid propellant engines could be heard rising from a 300-acre tract of wooded land in a Berlin suburb where the VfR had established a *Raketenflugplatz* (rocket flying field).

For von Braun and his fellow enthusiasts, the turning point came in the spring of 1932, when three German army officers visited the Raketenflugplatz. Intrigued by the potential of rocket weapons, the army offered to put selected members of the group to work in a government research laboratory. The experimenters jumped at the chance. "We were interested solely in exploring outer space," von Braun would explain years later. "We needed money."

Twenty years old, Wernher von Braun became technical head of the German military rocket program. The scope of the effort continued to expand after 1933, when Adolf Hitler and the Nazi party came to power. By 1936, the von Braun team had transferred to a fully equipped government rocket re-

*Von Braun (sitting on table) with the other members of the Army Ballistic Missile Agency at Huntsville, Alabama, in 1956.*

search and launch facility at Peenemünde, on the Baltic Sea. They spent the next eight years struggling to perfect the A-4, a rocket that would become better known by its military designation—V-2 (Vergeltungswaffe Zwei, or vengeance weapon two).

The work did not proceed smoothly. The members of the team had to overcome enormous technical problems while battling for continued government support. Hitler's enthusiasm for the project rose and fell like a roller coaster. The pres-

sure to achieve success grew much heavier with the beginning of World War II, and members of the team often faced real dangers. Von Braun himself was arrested by the German secret police in 1944, charged with intentionally retarding work on the V-2 and planning to defect to England with his rocket plans. General Walter Dornberger, the military commander of the Peenemünde facility, secured his release by insisting that the young scientist was essential to the success of the project.

The V-2, first successfully launched on October 3, 1942, was the most important single step on the road to space since Robert Goddard had sent the first liquid propellant rocket aloft in 1926. Standing more than 46 feet tall, it was guided to its target by gyroscopes in the nose controlling movable vanes that interacted with the rocket exhaust. The engine delivered an average 56,000 pounds of thrust at sea level, enough to boost the 27,000-pound missile to an altitude of 50 miles and giving it a range of 120 miles.

On the evening of the first successful launch, General Dornberger offered a toast to the scientists and technicians of Peenemünde. "Today," he remarked, "the spaceship was born!" If so, von Braun noted, "it landed on the wrong planet." Far from being a spaceship, the V-2 was a guided missile specifically designed to deliver nearly one ton of high explosives to the heart of London.

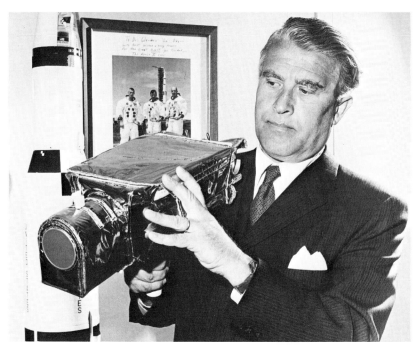

In 1971, von Braun examines a television camera for NASA's fourth manned lunar landing mission, Apollo 15.

From September 1944 to March 1945, as many as 100 V-2s fell on southern England, taking some 2,500 lives. And that was only the beginning of the human toll exacted by the rocket. With major German industrial centers undergoing round-the-clock bombing by the fall of 1944, the V-2s were assembled under the most difficult conditions by slave laborers who were literally worked to death.

With the German Reich collapsing around their ears and the victorious Red Army approaching Peenemünde in March 1945, von Braun and the key members of his team loaded their precious records into trucks and fled west in order to surrender to American troops. "The reason that I chose America in particular," he remarked a few years later, "is that America had a reputation for having an especially intense devotion to individual freedom and human rights." On another, less guarded, occasion he simply remarked that "the next time, I wanted to be on the winning side."

Following an intensive interrogation by both U.S. and British intelligence officers, von Braun and 127 Peenemünde colleagues were offered one-year contracts as employees of the American government. They were transported to Fort Bliss, Texas, where they helped to reassemble captured V-2 rockets and train the American crews who would test-fire the missiles at the White Sands Proving Ground, near Alamogordo, New Mexico.

The members of the team remained in America. Von Braun returned to Germany to wed his childhood sweetheart in 1947 and became a U.S. citizen in 1955. Although they did train American technicians and engineers, the Germans from Peenemünde retained leadership in rocket technology, eventually rising to positions of importance in NASA, the military space programs, and the aerospace industry. And no one was more important, or more visible, than Wernher von Braun.

During the lean postwar years, when congressional appropriations for rocket research were very low, von Braun never lost an opportunity to publicize the dream of spaceflight. He wrote popular magazine articles and books on the subject and made repeated appearances on television. By the mid-1950s, his efforts had convinced a generation of young Americans that the space age was just around the corner.

Along with many other members of his team, von Braun moved to Huntsville, Alabama, in 1950, where he served first as chief of development for the Guided Missile Division of the Army's Redstone Arsenal and then, after 1956, as director of development for the Army Ballistic Missile Agency. In that capacity he was responsible for the Jupiter, Juno, and Redstone rocket programs and played a key role in the launch of *Explorer 1*, the first U.S. artificial satellite. When NASA took over the army program in 1960, von Braun remained as the first director of the George C. Marshall Space Flight Center and supervised the development of the Saturn family of rockets that carried human beings to the moon.

The man who had joined the fledgling VfR almost half a century before was named deputy associate administrator of NASA in 1970. He served two years in that post before retiring from government service to become vice-president of engineering and development with Fairchild Industries, where he remained until his death on June 16, 1977.

Wernher von Braun was one of the great engineers of the 20th century. He surmounted overwhelming odds and exhibited courage, dedication, and genius in pursuit of what many regarded as an impossible goal. At the same time, his career illustrates the kind of ethical dilemmas that can surface when the quest for knowledge outweighs all other considerations.

Von Braun believed, above all things, in the dream of spaceflight and was willing to pay any price to make that dream come true. The entire world held its breath in wonder when his giant Saturn 5 boosted the *Apollo 11* astronauts off the pad and on their way to the moon. The men and women who died as slaves assembling V-2s in the caves at Nordhausen were not on hand to join in the applause.

*On January 31, 1958, Wernher von Braun and his colleagues launched Ex-*
*plorer 1, the first U.S. satellite, into orbit from Cape Canaveral, Florida,*
*aboard a Jupiter-C rocket. The satellite later made an important discovery of*
*radiation belts trapped in Earth's magnetic field.*

(*continued from page 45*)

Slowly but surely, NASA officials were setting the stage for a much more spectacular American space effort. By 1960, the agency had taken full control of the military programs originally transferred to it in 1958. In addition, the civilian space agency was gobbling up other key organizations and projects from the armed forces, including both the von Braun team of rocket specialists at the Development and Operations Division of the Army Ballistic Missile Agency and the deep space programs under way at the Jet Propulsion Laboratory of the California Institute of Technology.

In March 1961, NASA opened the first of its own new facilities, the Goddard Space Flight Center, in Greenbelt, Maryland. Goddard was to become the nerve center of the U.S. satellite effort, the place where earth-orbiting spacecraft were designed and built and from which they were monitored and controlled following launch from Cape Canaveral.

NASA engineers were now planning a new generation of space launch vehicles to carry those satellites into orbit. Scout, a solid propellant booster, was designed to loft relatively lightweight payloads into low earth orbit. Centaur, a high-energy upper stage fueled by liquid hydrogen, would boost robot probes into deep space, whereas the Saturn family of heavy launch vehicles might someday carry human beings to the moon.

Until those rockets were ready for flight, NASA would have to make do with its stable of Jupiter, Redstone, Thor, and Atlas missile/launch vehicles. The going was rocky at the outset. All four of NASA's attempts to launch payloads in 1958 ended in failure. The record for 1959 stood at 8 successes for 14 tries. In 1960, 17 launch attempts yielded 10 successes.

Still, the record was improving. The 18 successful launches of 1959–60 included such impressive achievements as the following:

- The first photo of the earth from space (*Vanguard 2*, February 7, 1959).
- The first successful weather satellite (*Tiros 1*, April 1, 1960).
- The first navigation satellite (*Transit 1B*, April 13, 1960).
- The first communications satellite (*Echo 1*, August 19, 1960).

Americans took pride in those accomplishments, but they reserved their real enthusiasm for the effort to send human beings into orbit. The military services had rushed forward with competing proposals for manned space programs immediately after the launch of *Sputnik 1*. The army called for a minimum suborbital effort, a quick trip up and back that Hugh Dryden of the NACA compared to "shooting a young lady out of a cannon." The navy proposed a much more complex scheme involving the return of the astronaut aboard an inflatable glider. Air Force officials promised to put Man In Space Soonest (MISS) with a single-seat spacecraft launched by a modified Atlas missile.

The best approach to manned spaceflight was also a matter of heated debate within the NACA. Traditionalists argued for a new generation of research aircraft capable of taking off from a runway, flying into space, and gliding back to a landing on earth. H. Julian Allen of the Ames Research Center observed that although such a plan might be the safest and surest way to send human beings into orbit, it was by no means the fastest. If the goal was to put Man In Space Soonest, engineers would have to abandon conventional aerodynamic thinking in favor of something like the army's "cannonball" scheme.

Allen suggested using an existing missile to launch a blunt, possibly saucer-shaped, spacecraft into orbit. He admitted that there was nothing fancy

*A Mercury space capsule is hoisted into position above a Redstone rocket in preparation for a flight test in 1959. The exterior of the cone-shaped craft contained protective metal shingles that were designed to deflect heat. Parachutes in the heat-resistant nose of the craft were designed to drop the capsule into the ocean after reentry.*

or elegant about this approach. "You just throw it," he remarked, then allow it to fall back to earth from orbit, relying on the blunt design to generate the air resistance that would slow its speed during reentry.

Fascinated by Allen's concept, Langley engineer Maxime Faget began to work out the details of the plan. He proposed a small, cone-shaped spacecraft, light enough to be launched on suborbital flights by a Jupiter-C rocket or into orbit by the more powerful Atlas missile. When the time came to return from space, a small cluster of rockets on the blunt end of the vehicle would be fired in the direction of motion to slow its speed and cause it to fall out of orbit. The craft would reenter the atmosphere blunt end first. Parachutes would lower it down to a landing in the ocean.

There were still problems to be resolved. The atmospheric friction that slowed the craft would also generate more than enough heat to destroy it. Faget was confident that such difficulties could be overcome. "As far as reentry and recovery are concerned," he noted, "the state of the art is sufficiently advanced to proceed confidently with a manned satellite project based upon the use of a blunt-body vehicle."

The broad outline of a plan to send human beings into space in a relatively short period of time was in place by July 29, 1958, when President Eisenhower signed the National Aeronautics and Space Act. Along with many other responsibilities, the act charged the new civilian space agency with ". . . the development and operation of vehicles capable of carrying instruments, equipment, supplies and living organisms through space." NASA, not the army, navy, or air force, would be sending the first Americans into orbit.

Several months before NASA opened its doors for business, NACA officials put Faget and a handful of colleagues to work transforming the Allen-Faget concept into the detailed design for a spacecraft. The vehicle that took shape on the drawing boards at Langley consisted of a titanium pressure vessel, just large enough to house a single occupant, surrounded by a conical outer shell of protective metal-alloy shingles designed to radiate heat away from the

*Maxime Faget, an engineer at Langley Research Center, worked out the details of the blunt-end design of the Mercury spacecraft. A special plastic and fiberglass shield had to be devised to protect the craft from the extreme temperatures it would be subjected to when it reentered Earth's atmosphere.*

spacecraft. A heat-resistant beryllium canister on the nose contained the parachutes that would drop the capsule into the ocean at the conclusion of the mission.

From the point of view of a prospective astronaut, the convex heat shield fitted to the blunt base of the craft was the most important single feature of the design. When subjected to extreme temperatures, the resins in the plastic and fiberglass shield would give off gases, dissipating heat and protecting the vehicle during its long plunge back into the atmosphere. A cluster of three solid propellant rockets strapped to the shield would be fired to initiate reentry, then jettisoned.

NASA was less than a week old when Administrator Glennan approved the Faget plan and created a Space Task Group (STG)—35 Langley engineers headed by Robert Gilruth—to push the work forward. On November 7, 1958, members of the STG released the specifications for their spacecraft and invited 40 aerospace firms to bid on its construction. NASA officials publicly announced the details of Project Mercury, the U.S. manned space program, on December 17, 1958, 55 years to the day since the Wright brothers had flown at Kitty Hawk. Less than a month later, on January 9, 1959, agency officials revealed that McDonnell Aircraft had won the bid competition and would serve as the prime contractor for the Mercury spacecraft.

The business of selecting the human beings who would ride the Mercury spacecraft into orbit was complete as well. The astronauts had passed through an extraordinary screening process. "What we're looking for," one air force general remarked, "is a group of ordinary Supermen."

At the outset, some federal officials had argued that the first space travelers should be chosen from the ranks of daredevils and athletes—race car drivers, high divers, or circus performers, men (no women were considered) accustomed to taking great risks. That did not seem appropriate to President Eisenhower, who decided that applicants would be selected from the existing pool of trained military test pilots.

An initial search revealed a total of 508 possible candidates. A study of personnel records, the comments of commanding officers, and personal interviews finally reduced that number to 32 test pilots. Each of these men was 40 years of age or younger, weighed less than 180 pounds, stood less than 5 feet 11 inches tall, and held an engineering degree. The finalists were probed, prodded, and subjected to batteries of psychological tests to determine which seven men would be chosen.

On April 9, 1959, NASA introduced the seven Mercury astronauts to the world. The group included three U.S. Air Force officers (Captains Donald K.

"Deke" Slayton, Virgil I. "Gus" Grissom, and L. Gordon Cooper), three naval aviators (Lieutenant Commanders Alan B. Shepard, Jr., Walter M. Schirra, Jr., and Lieutenant Malcolm Scott Carpenter), and a lone marine, Lieutenant Colonel John H. Glenn, Jr.

The astronauts were instant heroes. Like knights of old, they were seen as champions, men who would risk their life to carry their nation's banner in the race against the Soviets. Over the next few years, their names and faces would become familiar to all Americans. These were the men who would lay the doubts about America's strength and determination to rest and restore a sense of national pride and confidence.

Ironically, the precise duties of an astronaut were not entirely clear at the time of the selection. "Although the entire satellite operation will be possible without the presence of man," one NASA source reported, "the astronaut will play an important role." Small wonder that the astronauts should express some concern about their role on a mission that could be performed "without the presence of man," particularly when it was announced that chimpanzees would take the first rides into space aboard the Mercury capsule. The word within the test-flying fraternity was that the astronauts would be nothing more than "spam in the can," passengers who were simply going along for the ride.

Determined to ensure that they would function as pilots exercising a measure of control over the spacecraft, the astronauts used their growing prestige and public visibility to force essential design changes. At their insistence, tiny portholes gave way to larger windows; an emergency hatch was installed for quick escapes; and the location and function of every control, switch, lever, and instrument were studied and modified as required.

Repeated problems with the launch systems underscored the dangers of the enterprise. The Atlas rocket that would eventually boost the astronauts into orbit was plagued by developmental problems. Even the tried-and-true Redstone was still experiencing difficulties. The first complete flight test of the Mercury Redstone system at Cape Canaveral in November 1960 was an embarrassing flop. The rocket rose only six inches off the ground before an automatic engine shutdown settled it back onto the pad.

Nor did the first flight test of the Mercury Redstone combination with a living creature on board do much to boost astronaut confidence. Ham, a chimpanzee, made the suborbital trip into space on January 31, 1961. A series of malfunctions kept the vehicle and its occupant on the pad three hours longer than planned. Once aloft, the Jupiter-C subjected Ham to a much faster acceleration and higher gravitational forces than predicted. A leaky valve led to a severe drop in cabin pressure. The spacecraft splashed down 130 miles from

55

the target area. A damaged window allowed water to leak into the spacecraft while a safe but wet and badly frightened chimp waited for his delayed pickup.

With the whole world watching, safety was NASA's first priority. As a result, the first scheduled manned launch was shifted from March to late April 1961 to allow for changes and additional testing following Ham's nearly disastrous flight. As a result, on April 12, 1961, Major Yury Alekseyevich Gagarin, a 27-year-old Soviet air force pilot, became the first human being to fly in space. His vehicle, *Vostok* ("East"), weighed three times as much as a Mercury spacecraft. Moreover, the Soviet cosmonaut completed a full orbit of Earth, a much more impressive achievement than the suborbital trips planned for the first few manned Mercury flights.

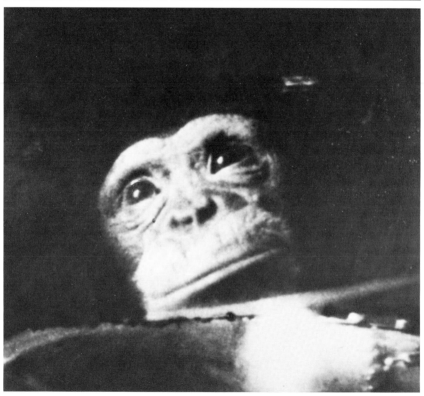

*Ham, the first chimpanzee to ride into space, is photographed aboard a Mercury capsule during the suborbital flight on January 31, 1961. Because of malfunctions during Ham's flight, NASA had to postpone the first scheduled manned launch so that its scientists could complete further testing.*

*Astronaut Alan B. Shepard, Jr., the first American in space, peers up at a U.S. Marine helicopter recovery team minutes after* Freedom 7's *splashdown on May 5, 1961. Shepard's successful suborbital flight lasted 15 minutes.*

Alan B. Shepard followed Gagarin into space on May 5, 1961. The Redstone booster lobbed Shepard's *Freedom 7* spacecraft 300 miles downrange at a speed of 5,000 miles per hour. At his maximum altitude of 100 miles, the astronaut exercised full control over the attitude of his craft. It was all over in 15 minutes.

Alan Shepard described his journey into space as "just a pleasant ride." But most Americans thought that it represented a great deal more than that. The Soviets had taken an early lead, but the space race was far from over. The flight of *Freedom 7* demonstrated that the United States was still very much in the running.

*On June 3, 1965,* Gemini 4 *astronaut Ed White "walks" in space for approximately 20 minutes while tethered to the spacecraft. White and fellow astronaut James A. McDivitt performed scientific and engineering experiments during the 62-revolution mission that lasted 4 days.*

# FIVE

# The Decision to Go to the Moon

On May 25, 1961, just 20 days after Alan Shepard rode *Freedom 7* into history, President John F. Kennedy addressed a joint session of Congress on the subject of urgent national needs. He concluded his talk with an extraordinary challenge. "I believe," he said, "that this nation should commit itself to achieving the goal, before this decade is out, of landing a man on the moon and returning him safely to earth."

John Kennedy's ringing call to send Americans to the moon came as something of a surprise to a great many observers. Before taking office, the new president had asked his science advisers to study the issue of manned spaceflight and suggest appropriate policies for the new administration to pursue. Like President Eisenhower's scientific advisory board before them, the Kennedy group argued for the development of scientific and applications satellites that would increase knowledge, revolutionize communications, and improve the accuracy of weather forecasts. They bluntly advised that "a crash program aimed at placing man into orbit at the earliest possible time cannot be justified . . . on scientific or technical grounds."

The attitudes of the Eisenhower and Kennedy scientific advisory panels underscored a basic tension within the U.S. space program. If the primary goals of the American effort were to acquire new scientific knowledge and to use the unique orbital vantage point to improve life on earth, then it was not really necessary to send human beings into space. Unmanned satellites and robot lunar and planetary explorers could accomplish those tasks at a fraction of the cost of a manned program.

Recognizing the enormous symbolic importance of matching the Soviets in space, Eisenhower had reluctantly overruled his advisers, approving NASA's Project Mercury and pressing for the immediate development of the Saturn family of very large boosters. President Kennedy reached the same decision with much less reluctance. A young and vigorous man, he came to office offering Americans the challenge of a "new frontier," but the first few months of his administration brought frustration and disappointment. A U.S.-backed invasion of Cuba ended in humiliating defeat. There was labor unrest at home

*President John F. Kennedy addresses a joint session of Congress on May 25, 1961, and challenges the nation to land men on the moon and return them safely to Earth before the end of the decade. Kennedy's call captured the public's imagination and Americans soon rallied around NASA's space program.*

and a crisis brewing in Southeast Asia. The new president seemed to come off second best in his first meetings with Premier Nikita Khrushchev of the Soviet Union and Charles de Gaulle of France.

John Kennedy saw the space race as a crusade behind which Americans could rally. Success would restore self-confidence at home, rebuild national prestige abroad, and allow the United States to stand up to the Soviets without risking nuclear confrontation. The problem was to choose a spectacular goal in space that America could be sure of attaining before the Soviets. Secretary of Defense Robert S. McNamara argued that the Soviet lead in space was so enormous that the United States could not be certain of winning anything short of a race to Mars. Vice-president Lyndon B. Johnson assured the president that a journey to the planets would not be necessary. NASA could beat the Soviets to the moon.

There was, of course, a drawback: A program to put human beings on the moon in less than a decade would be enormously expensive. Unwilling to fully advocate what might be seen as a waste of tax dollars, the president suggested the lunar journey as a tempting possibility. The final decision would depend on the response of the American people and the willingness of Congress to fund the great adventure.

The sheer excitement of the idea instantly captured the public imagination. Congress voted the funds, and the taxpayers applauded. There would be scattered voices of dissent as the price tag for Apollo rose to an estimated $22–$25 billion. Civil rights leaders and social activists questioned the value of a trip to the moon at a time when the nation faced serious domestic problems. Many scientists found it difficult to work up much enthusiasm for a program in which the journey itself seemed more important to mission planners than the new knowledge that would result. Caught up in the excitement of a drama complete with a little band of heroes facing cosmic dangers, few Americans were listening to the skeptics.

Perhaps more than any other single individual, James Edwin Webb deserves ultimate credit for the success of the American manned space effort from 1961 to 1968. Webb, whom Kennedy named to replace T. Keith Glennan as administrator of NASA early in 1961, was a professional manager with long experience in business and government. He had served in the State Department and headed the Bureau of the Budget during the Truman administration before entering private industry, where he rose to a position of leadership in the Kerr-McGee Company, a national leader in petroleum production and nuclear energy. As an ex-marine pilot and a member of the board of directors of McDonnell Aircraft, he was no stranger to aerospace.

President Kennedy, Congress, and the American people had decided to go to the moon. James Webb was just the man to get the job done. Webb was a technocrat—a man who believed that the power of a modern nation depended on its ability to manage large-scale technical enterprises. "The great issue of this age," he wrote, "is whether the U.S. can, within the framework of existing economic, social, and political institutions, organize its development and use of advanced technology as effectively for its goals as can the Soviet Union."

For Webb, the trip to the moon offered a golden opportunity to teach the world a lesson in management and administration. By successfully marshaling the forces of government, industry, and the universities to accomplish a monumental task, NASA would demonstrate how to make decisions, organize resources, and manage complex systems in a way that could be used to solve the social, economic, and political problems that plagued the world.

James Webb was a master politician, as well. He knew precisely how to win support for his program, as Robert Gilruth, the head of Project Mercury, learned during the planning for the Johnson Space Center. NASA, facing a period of enormous growth, required a new facility to house the manned space effort. When Webb announced that a new center would be constructed for that purpose in Houston, Texas, Gilruth suggested that it might be easier and less expensive to expand the existing Langley Research Center in Virginia. "Bob," Webb replied, "what the hell has Senator Harry Byrd [of Virginia] ever done for you or NASA?" Houston, he pointed out, was in Representative Albert Thomas's congressional district. Administrator Webb wanted to have Thomas, chairman of the House Appropriations Subcommittee controlling NASA's budget, in his corner.

The ability to win support for his program was only one of Webb's management skills. The trip to the moon was an enormously complex undertaking involving billions of dollars and millions of men and women in NASA centers and contractor facilities scattered from coast to coast. Every spacecraft and every launch vehicle would be made up of thousands of individual parts and pieces, each of which had to be manufactured, tested, and shipped precisely on schedule and within budget. There was no margin for error. The failure of the smallest part at a critical moment could spell disaster and death for a crew.

James Webb wanted to be certain that there were no misunderstandings within NASA. He and his headquarters staff were ultimately responsible for administering every phase of the program. They set the schedules, established the budgets, and monitored performance. Safety and quality assurance were their primary concerns. Webb had every confidence in the scientists and

*On February 14, 1961, James E. Webb (center) is sworn in as NASA admin-*
*istrator. Webb, an ex-marine pilot and a former head of the Bureau of the*
*Budget, was a technocrat who knew how to win congressional support for*
*NASA programs.*

engineers working at the various centers. Nevertheless, he insisted on maintaining an independent Office of Quality Assurance whose job it was to spot problems and report them directly to the administrator.

The business of planning a series of progressively more ambitious flights into space that would culminate in a landing on the moon was soon under way. Wernher von Braun and the launch vehicle experts at what was now the Marshall Space Center in Huntsville, Alabama, originally argued for the use of an enormous 600-foot-tall "dream rocket" called Nova. The giant booster

*In July 1969, the 363-foot-high Apollo 11 Saturn 5 spacecraft undergoes tests while on launchpad 39A at Kennedy Space Center. The 3-stage Saturn 5 rocket, capable of lifting the weight of 120 Mercury spacecraft into orbit, had to be assembled on a mobile launch platform in a specially designed building and then trucked, standing straight up, to the launchpad.*

would be capable of sending three astronauts directly to the moon in a spacecraft large and powerful enough to land the entire crew on the lunar surface and return them directly to Earth.

Bob Gilruth and his Mercury team disagreed. They proposed using a large rocket already on the drawing boards, the Saturn C-5 (renamed Saturn 5 in 1963), to send two much smaller spacecraft modules into lunar orbit. One part of the system, the Command and Service Module (CSM), would remain in orbit with a single crew member on board, while the other two astronauts descended to the surface in a Lunar Module (LM). A small on-board rocket motor would return the LM and its occupants back into lunar orbit, where they would rendezvous and dock with the CSM. The larger rockets of the CSM would then take the combined spacecraft out of lunar orbit and return them to Earth.

Gilruth's plan sounded complicated and risky, but it had several great advantages. The Saturn C-5 would be an enormous rocket, standing 30 stories

64

tall and capable of lifting the weight of 120 Mercury spacecraft into orbit, but it was much smaller and simpler than the Nova. Eventually, even von Braun came to realize that the orbital rendezvous approach would get men to the moon much quicker and less expensively than his own plan.

The course was now clear. NASA would move toward a lunar landing through a series of flights with three types of spacecraft. Project Mercury would be followed by additional earth-orbital missions flown with an enlarged two-man Gemini version of the basic spacecraft design. The Gemini program would test advanced equipment, give the astronauts their first opportunity to venture outside the spacecraft, and practice the rendezvous and docking maneuvers that would be required for the moon mission. With all of that accomplished, the way would be open for Project Apollo, which would culminate in a series of landings on the lunar surface.

While the decision making was taking place behind the scenes, public attention focused on the flights themselves. The Mercury, Gemini, and Apollo

*In the early 1960s, Project Mercury astronauts show the design of the Atlas booster and Mercury capsule. The astronauts are (seated, left to right) Gus Grissom, Scott Carpenter, Deke Slayton, L. Gordon Cooper, and (standing, left to right) Alan B. Shepard, Jr., Wally Schirra, and John Glenn.*

programs provided more than enough excitement to keep everyone interested. Virgil I. "Gus" Grissom followed Alan Shepard into space aboard the Mercury spacecraft *Liberty Bell 7*. Grissom was bobbing safely around in the Atlantic at the conclusion of his suborbital flight on July 21, 1961, when the emergency escape hatch accidentally blew off, forcing him to make a hasty exit as water poured into the cockpit. The astronaut was rescued, drenched and exhausted, but his spacecraft was lost.

An Atlas rocket boosted John Glenn off the pad at 9:47 on the morning of February 20, 1962. Everything seemed to be going well on the first U.S. orbital mission until telemetry signals from the spacecraft suggested to ground controllers that the all-important heat shield of Glenn's *Friendship 7* had come loose. The entire nation seemed to hold its breath as Glenn dropped back toward earth and disappeared into a normal radio blackout that was part of the

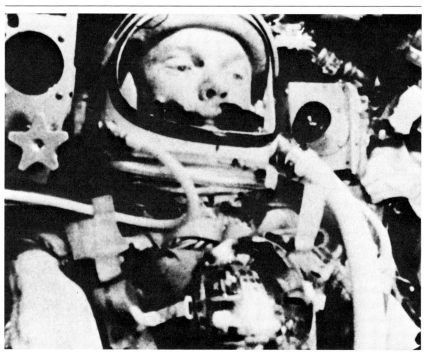

*On February 20, 1962, John Glenn became the first American to orbit Earth. Aboard Mercury's* Friendship 7 *for 4 hours and 55 minutes, Glenn eased fears about man's ability to endure weightlessness and buoyed the pride of his fellow astronauts when he had to manually take charge of the spacecraft after the automatic controls malfunctioned.*

reentry process. The heat shield held, and astronaut John Herschel Glenn emerged as one of the great heroes of the space program.

Six of the original seven astronauts flew into space during the two years between May 5, 1961, and May 15, 1963. Donald K. "Deke" Slayton, grounded by a minor heart irregularity, would eventually travel into space 10 years later as a member of the Apollo-Soyuz Test Project crew.

| Project Mercury Manned Mission | | | |
| --- | --- | --- | --- |
| Freedom 7 | May 5, 1961 | Alan Shepard | Suborbital |
| Liberty Bell 7 | July 21, 1961 | Virgil Grissom | Suborbital |
| Friendship 7 | Feb. 20, 1962 | John Glenn | 3 orbits |
| Aurora 7 | May 24, 1962 | M. Scott Carpenter | 3 orbits |
| Sigma 7 | Oct. 3, 1962 | Walter Schirra | 6 orbits |
| Faith 7 | May 15–16, 1963 | L. Gordon Cooper | 22 Orbits |

The Gemini program was the next giant step on the road to the moon. The two-man astronaut teams who flew the new spacecraft would find their quarters very cramped. Although the Gemini capsule was twice the size of the old Mercury spacecraft, every nook and cranny was stuffed with advanced electronic gear. On-board computers and radar systems enabled the crew to maneuver in space and change orbits to rendezvous and dock with another spacecraft, something that was impossible with Mercury. In addition, the Gemini spacecraft came equipped with life-support systems and fuel cells to produce enough water and electricity to sustain the crew on missions of up to two weeks in earth orbit.

The Gemini team experienced its share of difficulties. The spacecraft, too large for launch with an Atlas rocket, would be sent aloft aboard a modified version of the new Titan intercontinental ballistic missile being developed by the U.S. Air Force. Rising booster costs, slipping production schedules, and quality control problems with the Titan plagued mission planners who were also struggling to overcome the normal difficulties involved in the design and construction of a new spacecraft.

Despite these problems, the 10 manned Gemini flights were well worth the effort. One by one the techniques and equipment that would enable astronauts to journey to the moon were tested in space. There were also moments of high excitement, as when Ed White, one of the new astronauts recruited for the Gemini program, became the first American to "walk" in space on June 3, 1965.

There were dangers, as well. The first planned close approach of two U.S. crews in space began on December 4, 1965, when James Lovell and Frank

Gemini 6 *(foreground) and* Gemini 7 *rendezvous in space on December 15, 1965. The two spacecraft flew together for almost 5 hours, before* Gemini 7 *went on to complete its 14-day mission to determine the effects of prolonged space travel on the astronauts.*

Borman took off on *Gemini 7*. Just a week later, Walter Schirra and Thomas Stafford, the crew of *Gemini 6-A*, were poised on the pad at Cape Kennedy, ready to follow them into space. The Titan booster ignited, then immediately shut down, leaving the two astronauts trapped aboard a live rocket that might explode at any moment. Rather than activating the escape system, a procedure that would require canceling the dual mission, the two astronauts remained aboard the spacecraft until rescued. The gamble paid off. Three days later, following a flawless launch, they maneuvered their spacecraft to within inches of their companions aboard *Gemini 7*.

Neil Armstrong and David Scott had their own brush with disaster during the *Gemini 8* mission. Just after completing the first-ever docking in space with a specially equipped, unmanned Agena upper stage, their spacecraft began to roll out of control. Armstrong immediately cut loose from the Agena, only to find the situation growing worse. As they would later discover, a valve in the orbital maneuvering system was stuck. Fighting to maintain control, the crew was forced to make an immediate reentry. Neil Armstrong's piloting skills saved the day.

By almost any measure, Gemini was an enormous success. The NASA team overcame all of the obstacles to achieve 10 flights that went off like clockwork—1 every 2 months—and attained all of the goals established for the program. As Gemini drew to a close, Americans could look back on the country's recent accomplishments in space with considerable pride and look forward to a future that would include a landing on the moon, as John Kennedy had promised, before the decade of the 1960s was out.

| Project Gemini Manned Mission | | |
|---|---|---|
| Gemini 3 | March 23, 1965 | V. Grissom, J. Young |
| Gemini 4 | June 3–7, 1965 | J. McDivitt, E. White |
| Gemini 5 | Aug. 21–29, 1965 | L. G. Cooper, C. Conrad |
| Gemini 7 | Dec. 4–18, 1965 | F. Borman, J. Lovell |
| Gemini 6-A | Dec.15–16, 1965 | W. Schirra, T. Stafford |
| Gemini 8 | March 16, 1966 | N. Armstrong, D. Scott |
| Gemini 9-A | June 1–3, 1966 | T. Stafford, E. Cernan |
| Gemini 10 | July 18–21, 1966 | J. Young, M. Collins |
| Gemini 11 | Sept. 12–15, 1966 | C. Conrad, R. Gordon |
| Gemini 12 | Nov. 11–15, 1966 | J. Lovell, E. Aldrin |

*Astronaut Buzz Aldrin descends the ladder of the Lunar Module* Eagle *after landing on the moon on July 20, 1969. After more than 21 hours on the surface, the Lunar Module—a 2-man spacecraft—carried the astronauts back to the orbiting Command Module* Columbia.

# SIX

# A Chariot for Apollo

There never was a space race, at least not as most Americans conceived it. Soviet leaders had chosen space as the arena in which to demonstrate their technological prowess and test the resolve of a cold war rival, but the notion of racing the Americans toward the spectacular goal of a manned lunar landing was a very different matter.

The Soviets piled up a few more space firsts after 1966, but they could not match the rapidly developing technical sophistication of a U.S. program that had shifted into high gear. They fell behind in electronics; were slow to develop the capability to change orbits and dock; suffered the first fatal accident in space on April 23, 1967, when the parachute of cosmonaut Vladimir Komarov's Soyuz capsule failed to open; and experienced catastrophic difficulties with their large booster program. Recognizing that a flight to the moon was not a serious possibility for them in the immediate future, the Soviets chose to concentrate on a long-term program of manned operations in earth orbit.

With or without the spur of competition, American planners had no intention of stopping short of the final goal. Mercury and Gemini had set the stage for a lunar landing. The funding was in place, the hardware was under construction, and the crews were in training. People around the globe were waiting for the moment when the first human being would step onto the surface of another world.

Disaster struck when it was least expected, not with a crew halfway to the moon but during a routine ground test at the very outset of the program. Just after lunch on the afternoon of January 27, 1967, astronauts Virgil Grissom, Edward White, and Roger Chaffee, the first crew scheduled to fly Apollo into space, took the elevator to the top of a service tower at Cape Kennedy, walked through the "White Room" connecting the tower to the spacecraft, and climbed into the Apollo 204 Command Module. Five hours later they were still sealed in the spacecraft, practicing flight procedures while struggling to overcome a communications problem.

At precisely 6:31 P.M. an alarmed Gus Grissom called out over the radio: "There's a fire in here!" Smoke, heat, and fumes erupted into the White Room, preventing technicians from approaching the capsule. A team of workmen had the fire under control and the hatches open within five minutes, but it was too late. The three astronauts were dead of asphyxiation and burns.

The Apollo 204 catastrophe was not a freak accident. In the aftermath of the fire it was difficult to imagine how the danger could have been overlooked. The life-support system supplied an oxygen-rich high-pressure environment to a capsule filled with flammable materials and hundreds of potential sources of ignition. The new spacecraft had been the proverbial accident waiting to happen.

In fact, the dangers had not gone unnoticed. General Sam Phillips, an air force officer assigned to assist NASA with Apollo planning, had issued a series of stinging reports complaining of poor workmanship on the part of several contractors. An employee of North American Rockwell, the firm that built the Apollo CSM, had been fired for calling attention to safety problems related to the spacecraft. Just three weeks before the fire, astronaut physician Dr. Charles Berry complained of the specific hazards faced during ground tests, when the cabin pressure was much higher than it would be in space and the danger of fire therefore much greater.

Suddenly, the very future of Apollo was in danger. Without the presence of James Webb, working to restore confidence and get the program back on track, Project Apollo might have come to an end. Webb's first step was to establish a NASA accident investigation board, headed by Floyd Thompson, director of the Langley center. He then turned to face Congress, the White House, and the public, explaining and defending his program and justifying the allocation of an additional half a billion dollars with which to redesign and rebuild the Apollo spacecraft.

Webb knew what had occurred and what had to be done. He realized that the increasing pressure to move forward, coupled with the incredible complexity of

On January 27, 1967, the charred Apollo 204 spacecraft sits at Cape Kennedy following an electrical fire in which astronauts Gus Grissom, Ed White, and Roger Chaffee were killed. After the catastrophe NASA established an investigation board to reassess the program and to determine potential design problems in the spacecraft.

the hardware required to transport human beings to the moon, had forced safety and quality assurance into the background. That could never be allowed to happen again. Contractors were brought into line, new safety procedures instituted, and NASA employees forced to remain absolutely current on the status of every aspect of spacecraft and launcher development.

At the same time, the Apollo spacecraft was checked, rechecked, and redesigned. NASA investigators compiled an initial list of 8,000 potential problems that had to be resolved. Ultimately, 1,697 changes were recommended to the NASA Configuration Control Board established to make the final design decisions. The board approved a total of 1,341 alterations.

The bitter lessons of the Apollo 204 disaster would eventually be forgotten, but not before men reached the moon. The period of thoughtful reassessment and redesign following the fire gave hard-pressed engineers some welcome

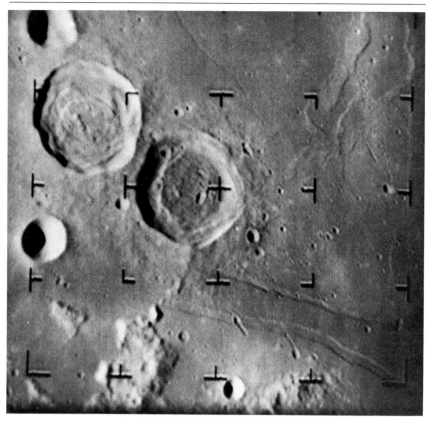

*A television picture taken by* Ranger 8 *on February 20, 1965, shows the large flat-bottomed craters of the moon. The first phase of the Apollo lunar landings involved a series of nine unmanned Ranger spacecraft designed to crash into the lunar surface, relaying photos back to earth right up to the moment of impact.*

breathing room. The pressure for immediate flights was relaxed. There was time to catch up with a program that had been racing ahead so rapidly that no one could keep track of all the details.

The unmanned exploration of the moon in preparation for eventual Apollo landings was also moving ahead. The first phase of the program involved a series of nine Ranger spacecraft designed to crash into the lunar surface, sending photos back to earth right up to the moment of impact. Ranger was a hard-luck program, plagued by personnel difficulties (that were intensified by the uneasy relationship between NASA and the Jet Propulsion Laboratory, the contractor-operated center that managed the program) and by hardware failures. The first 6 launches were unsuccessful, but Rangers 7, 8, and 9 performed flawlessly, providing much better images of potential landing spots than could have been obtained in any other way.

The second phase of the unmanned exploration of the moon was achieved as NASA engineers landed five Surveyor spacecraft safely on the lunar surface between May 1966 and January 1968. In addition to proving that the lunar surface would support the weight of a spacecraft, the Surveyors returned thousands of images of the lunar landscape and sampled and analyzed the soil. In conjunction with the Surveyors, five Lunar Orbiter spacecraft were launched from August 1966 to August 1967. These spacecraft flew a total of 6,000 orbits around the equator of the moon, photographing 99 percent of its surface. At the conclusion of their missions, each of the Orbiters was commanded to crash on the moon to avoid endangering later manned operations.

Within 10 months of the tragic loss of Grissom, White, and Chaffee, NASA was back in the headlines with another triumph. On November 9, 1967, the first full-scale Saturn 5, complete with an empty Apollo Command and Service Module, roared aloft from Pad 39 at the Kennedy Space Center. Astronaut Michael Collins, who would ride another Saturn 5 to the moon less than two years later, was watching from four miles away. It was a moment he would never forget.

> At ignition the flame was orange-red, but rapidly changed to an incandescent white at its core and a dirty brown at its edges. The scene had an eerie quality because for the first 20 seconds it occurred in total silence. When the sound wave reached us with a sudden jolt, it was more than just a noise. The sand under my feet began vibrating and I felt as if a giant had grabbed my shirtfront and started shaking.

Everything about the Saturn 5 staggered the imagination. The entire rocket stack, from the base of the F-1 engine nozzles to the tip of the escape tower, stood 363 feet tall and weighed 6,100,000 pounds. The doors of the enormous Vehicle Assembly Building in which the Saturn 5s were stacked for flight were 48 stories high, large enough so that the United Nations building could have been wheeled through them.

The cluster of 5 F-1 engines that powered the first stage of the Saturn 5 burned kerosene fuel and liquid oxygen at the incredible rate of 3,333 gallons per second. The engines delivered a total of 7.75 million pounds of thrust, roughly the equivalent of 180 million horsepower. It is estimated that 2 million horsepower was converted to sheer noise. Little wonder that the launch of a Saturn 5 was something to be felt as well as heard.

Everything about the Apollo program was big, including the price tag. Each Lunar Module, the 2-man spacecraft that carried astronauts down to the moon, cost 15 times its weight in gold. Having come so far, Congress and the president were willing to pay an extraordinary price to see the American flag planted on the lunar surface. It was clear, however, that there would be little support for major new space projects after Apollo.

As a senator and vice-president, Lyndon Johnson had been one of the most important political supporters of the U.S. space effort. As president, his priorities shifted. Determined to fund a series of major social programs aimed at building a "Great Society" while fighting an enormously expensive and ever-expanding war in Southeast Asia, Johnson could no longer justify supporting NASA at the level of the peak Apollo years.

The NASA budget increased every year from 1959 to 1965. In 1966, for the first time in the history of the agency, the budget figure was lower than that of the year before. It was the beginning of a steady decline that would continue for 10 years. The number of NASA employees peaked at 35,860 in 1967 and has declined every year since then. The flush times were over for the nation's civilian space agency.

James Webb put NASA back on the road to the moon, but he did not stay to enjoy the moment of triumph. A Kennedy appointee, he had enjoyed a friendly but not particularly close relationship with President Johnson, who had installed one of his own appointees, Thomas O. Paine, as deputy administrator. During a visit to the White House in October 1968, Webb suggested that he might ask to retire in the not too distant future. He was stunned when the president accepted his "resignation" on the spot.

Twenty-one years after he left office, James Edwin Webb remains perhaps the single most important figure in the history of NASA. When he came to the agency in 1961, it was a relatively small organization, struggling to keep up with the Soviets in space. During the seven years of his administration, he transformed NASA into one of the largest and most visible agencies of the federal government. Webb, the technocrat, succeeded in uniting the forces of government and industry to achieve a great national goal. At its peak, the civilian space agency and its contractors employed 430,000 people—from astronauts and engineers to animal keepers, dietitians, and filmmakers. In the process, the agency funded research and development in the most advanced areas of technology: computers and electronics, energy, metallurgy and

*On October 22, 1968, aboard the USS* Essex, Apollo 7 *astronauts (left to right) Donn F. Eisele, Wally Schirra, and R. Walter Cunningham speak with President Lyndon B. Johnson after their 11-day earth-orbital mission. The astronauts completed 163 orbits and beamed the first television pictures to Earth from a manned spacecraft.*

materials processing, and chemistry, to name but a few. Webb had achieved his goal. Under his leadership, NASA had shown the world how to manage a vast technological enterprise.

Thomas Paine took over as administrator of an agency almost fully recovered from the effects of the Apollo 204 fire. The first of the redesigned Apollo Command Modules arrived at Cape Kennedy in May 1968 with a message from North American Rockwell stenciled on the crate: "We care enough to send the very best." The crew of *Apollo 7* (Grissom, White, and Chaffee counted as *Apollo 1*, followed by 5 unmanned launches) flew the CSM into earth orbit on October 11. Astronauts Walter M. Schirra (the only man to fly in the Mercury, Gemini, and Apollo programs), Donn F. Eisele, and R. Walter Cunningham remained in space for 10 days and 20 hours, completing 163 earth orbits, beaming the first television pictures to earth from a manned spacecraft, and turning in a nearly flawless performance.

Apollo 8 *astronaut Frank Borman waves good-bye before TV transmission ends for the day on December 22, 1968. The astronauts of* Apollo 8 *were the first to orbit the moon.*

*Apollo 8* represented a great step forward for the program. Astronauts Frank Borman, James Lovell, and William Anders set out on the first voyage to the moon on December 21, 1968. The Lunar Module was not yet ready to fly, so there would be no descent to the surface. The crew spent 20 hours in lunar orbit. The high point of the mission, in many ways the high point of the entire U.S. manned space program, came on Christmas Eve, 1968, when millions of people around the globe watched a special television broadcast from three astronauts circling the moon. The crew of *Apollo 8* described their mission and offered viewers a glimpse of the lunar surface sweeping past the window of the spacecraft, after which mission commander Frank Borman read from the Book of Genesis and sent holiday greetings back to everyone "on the good Earth."

Not since the flight of John Glenn, almost seven years before, had American enthusiasm for the space program reached such a peak of excitement. Human beings had yet to step on the moon, but they had journeyed there and returned home safely. The enduring legacy of *Apollo 8* is to be found in an image of home, a lovely blue-green sphere suspended above the harsh lunar landscape. Set against the blackness of space, the earth looks small, fragile, and incredibly inviting. Seldom has a single photograph had so profound an impact, reminding everyone of the fragility of our planet and of the need for cooperation among all of those who call it home.

Two more manned missions followed in quick succession. The crew of *Apollo 9* checked out all of the hardware, including the Lunar Module, in earth orbit. *Apollo 10* went back to the moon, where mission commander Thomas Stafford and Eugene Cernan flew the Lunar Module to within 50,000 feet of the surface and returned to lunar orbit for a safe docking with the Command Module. Just eight years before, President John Kennedy had challenged the nation to journey to the moon before the end of the decade. Now the stage was set for the great event.

*A Saturn 5 rocket lifts off at 9:32 A.M. on July 16, 1969, carrying the crew of* Apollo 11. Apollo 11, *the first manned lunar landing mission, was the climax of NASA's Apollo program.*

79

The crew of *Apollo 11*, mission commander Neil Armstrong, Lunar Module pilot Edwin "Buzz" Aldrin, and Command Module pilot Michael Collins, launched from Pad 39A at Cape Kennedy on July 16, 1969. Three days later they entered lunar orbit. "Despite the fact that I have spent years studying photographs from Ranger, Lunar Orbiter and Surveyor, as well as from Apollo 8 and 10," Mike Collins later remarked, "it is nevertheless a shock to actually see the moon at first hand."

> The first thing that springs to mind is the vivid contrast between the earth and the moon. . . . I'm sure that to a geologist the moon is a fascinating place, but this monotonous rock pile, this withered, sun-seared peach pit out my window offers absolutely no competition to . . . the earth, with its verdant valleys, its misty waterfalls. . . . I'd just like to get our job done and get out of here.

The drama built toward a climax as Armstrong and Aldrin descended toward the surface in the Lunar Module *Eagle*. Discovering that the site selected for the landing was strewn with boulders, Armstrong took control of the spacecraft

*On July 20, 1969, Lunar Module pilot Buzz Aldrin poses for a portrait; the Lunar Module and mission commander-photographer Neil Armstrong can be seen in the reflection on Aldrin's visor. Between July 1969 and December 1972, NASA launched six successful missions to the moon.*

*An* Apollo 11 *astronaut's footprint lies in the lunar soil. While on the moon the astronauts collected soil and rock samples, conducted solar wind experiments, and set up seismic equipment to record moonquakes.*

away from the computer, maneuvering across the surface before finally setting down in an open area of the Sea of Tranquility with 30 seconds' worth of fuel to spare. "Houston," he reported, "Tranquility Base here. The Eagle has landed."

Six hours later, on July 20, Neil Armstrong stepped onto the moon. The two astronauts surveyed the scene, planted the American flag, and spoke to President Richard Nixon while on the surface. The rest of the mission—lift-off from the moon, rendezvous with the Command Module in lunar orbit, jettisoning the Lunar Module and the return to earth, reentry and splashdown—went precisely according to plan. The first manned flight to the moon was a textbook mission.

*Apollo 11* was the first of six successful NASA expeditions to the moon between July 1969 and December 1972. Twelve astronauts would actually walk on its surface. They probed, poked, and sampled the lunar soil; photographed everything in sight; drove across the dusty plains in their Lunar Roving Vehicles; and returned a total of 838 pounds of moon rock to the earth for study.

| Apollo Manned Missions | | | |
|---|---|---|---|
| Apollo 7 | Oct. 11–12, 1968 | Walter H. Schirra<br>Donn Eisele<br>R. W. Cunningham | 10 days, 20 hours;<br>earth orbit |
| Apollo 8 | Dec. 21–27, 1968 | Frank Borman<br>James Lovell<br>W. A. Anders | 6 days, 3 hours;<br>lunar orbit |
| Apollo 9 | March 3–13, 1969 | J. McDivitt<br>David Scott<br>R. L. Schweickart | 10 days, 1 hour;<br>152 earth orbits |
| Apollo 10 | May 18–26, 1969 | E. Cernan<br>John W. Young<br>T. Stafford | 8 days, 3 minutes;<br>lunar orbit |
| Apollo 11 | July 16–24, 1969 | Neil Armstrong<br>E. Aldrin<br>M. Collins | 8 days, 3 hours;<br>first lunar landing |
| Apollo 12 | Nov. 14–24, 1969 | C. Conrad<br>R. Gordon<br>A. L. Bean | 10 days, 4 hours;<br>lunar landing |
| Apollo 13 | April 11–17, 1970 | J. Lovell<br>J. L. Swigert<br>F. W. Haise | 5 days, 22.9 hours;<br>lunar landing mis-<br>sion aborted as a<br>result of accident |
| Apollo 14 | Jan. 31–Feb. 9, 1971 | A. Shepard<br>S. A. Roosa<br>E. D. Mitchell | 9 days; lunar<br>landing |
| Apollo 15 | July 26–Aug. 11, 1971 | D. R. Scott<br>J. B. Irwin<br>A. M. Worden | 12 days, 17 hours;<br>lunar landing; first<br>use of the Lunar<br>Roving Vehicle |
| Apollo 16 | April 16–27, 1972 | J. W. Young<br>T. Mattingly<br>C. Duke | 11 days, 1 hour;<br>lunar landing |
| Apollo 17 | Dec. 7–19, 1972 | E. Cernan<br>R. B. Evans<br>H. H. Schmitt | 12 days, 13 hours;<br>lunar landing |

The Apollo voyages to the moon were completed by the end of 1972. The value of the program has been endlessly debated during the years since that time and will continue to be argued in the future. Was the all-out effort to put men on the moon worth the enormous expenditure of time, money, and effort? Would a slower, more cautious long-range manned space program with more modest goals have provided a greater scientific return or laid a firmer foundation for future space ventures? There are no simple answers to those and other questions.

Apollo 11 *astronauts Neil Armstrong, Michael Collins, and Buzz Aldrin await recovery by helicopter after they splashed down in the Pacific Ocean on July 24, 1969. The astronauts are wearing biological isolation garments and will be quarantined in a trailer aboard the USS* Hornet *while scientists study their possible contamination from lunar microbes.*

We can be certain of one thing, however. The men and women of NASA had accomplished the goal set for them 11 years before. Human beings had journeyed to the moon, explored its surface, and returned safely to earth. Whatever its practical benefits, Project Apollo was a triumph of the human will and spirit. There was no mistaking the ultimate message of the program. If human beings could fly to the moon, was there anything they could not accomplish?

83

*A model of the Voyager spacecraft undergoes tests at the Kennedy Space Center in 1977.* Voyager 1 *and* Voyager 2, *developed for NASA by the Jet Propulsion Laboratory based on the designs of the Viking and Mariner spacecraft, were launched to study Jupiter, Saturn, Uranus, and Neptune.*

# SEVEN

# Putting Space to Work

The National Aeronautics and Space Administration celebrated its 25th anniversary in 1983. The agency commemorated the occasion with museum exhibitions, films, and special publications. In all of the excitement, a genuine milestone passed almost unnoticed. At precisely 5:00 A.M. Pacific time, on June 13, 1983, *Pioneer 10* officially became the first product of human civilization to leave the Solar System.

Launched 11 years earlier, on March 3, 1972, *Pioneer 10* had already piled up an impressive list of firsts. It was the first spacecraft to travel beyond the orbit of Mars, to the outer planets, returning the first spectacular close-up images of Jupiter. Its primary mission complete, *Pioneer 10* passed out of the Solar System and into the infinity of interstellar space. NASA provided humanity's first messenger to the stars with a suitable calling card, a small metal plaque showing a man and a woman, a diagram of the Solar System, and a map indicating the position of the sun in the Galaxy.

While public attention was focused on the dramatic activities of the astronauts and cosmonauts, the men and women who design, build, and fly satellites and planetary probes were laying the foundations for a genuine space-flight revolution. They seldom made headlines, but their work touched the everyday life of people around the world in some very down-to-earth ways, while enabling scientists to probe the deepest mysteries of the universe.

*On October 19, 1983, President Ronald Reagan (center) addresses NASA employees during NASA's 25th anniversary celebration. NASA's first 25 years saw the completion of the Mercury, Gemini, Apollo, and Skylab projects and the beginning of the Space Shuttle program.*

NASA's applications satellite program, for example, is one of the great success stories of the space age. *TIROS 1* (Television and Infrared Observation Satellite), the world's first meteorological satellite, was launched on April 1, 1960. *Transit 1B*, the first navigation satellite, roared aloft from Cape Canaveral 12 days later. *Echo 1*, the world's first passive communications satellite, followed on August 12, 1960. These three satellites inaugurated the era of space applications.

NASA's role would be to blaze new trails, conducting the first experiments to demonstrate the value of a new technology and cooperating with other agencies and private corporations interested in developing even more sophisticated spacecraft to perform a specific task. With that accomplished, NASA would turn things over to an organization that would create and maintain an operational system.

# Communications Satellites

The growth of communications satellite technology established the pattern. *Echo 1*, NASA's first 30-foot inflatable communications satellite (comsat), was a passive target that simply reflected radio signals aimed at it from earth. *Courier*, the first active comsat capable of receiving, amplifying, and rebroadcasting messages from earth, was launched by the U.S. Army on October 4, 1960. NASA launched *Telstar 1*, the first active civilian comsat, developed by the American Telephone and Telegraph Company (AT&T), on July 10, 1962. The Radio Corporation of America's (RCA) *Relay 1*, the third active comsat, followed on December 13, 1962.

The Relay and Telstar series were technology demonstrators, transmitting the first civilian television broadcasts from America to Europe and Japan, but they flew at relatively low altitudes and had to be constantly tracked in orbit. It remained for the Hughes Aircraft Corporation to develop *Syncom 1* and *Syncom 2*, launched by NASA on February 14 and July 26, 1963. Orbiting 22,230 miles above the equator, they were geosynchronous satellites, meaning that their speed in orbit precisely matched the speed of the rotation of the earth, so that they remained positioned over one spot on the ground. Three such satellites placed equidistantly around Earth would provide continuous television and communications coverage to virtually the entire globe.

The success of the geosynchronous satellites demonstrated the commercial potential of a satellite-based global communications system. But how could the transfer of this new technology from government to private enterprise be accomplished? The Kennedy administration had the answer. On August 31, 1962, after an extended congressional debate, President Kennedy signed a bill creating the Communications Satellite Corporation (COMSAT), a unique organization that would be half-owned and controlled by the government and half by large communications companies willing to invest in the venture. The new firm would contract with aerospace companies for the design and construction of comsats and pay NASA to launch them. COMSAT Corporation would profit from fees paid for the use of its ground facilities and satellite channels.

COMSAT flourished. International agreements led to the creation of the International Telecommunications Satellite Consortium (Intelsat), of which COMSAT became the American member. NASA launched the organization's first satellite, *Early Bird*, on April 6, 1965. By 1969, membership in Intelsat had grown to 40 nations operating a successful and enormously profitable worldwide communications satellite network. The most important revolution in communications since the invention of radio was well under way.

NASA has continued to build and fly specialized communications satellites. Three of the six Applications Technology Satellites (ATS) launched between December 1966 and May 1974 explored new facets of communications satellite technology. *ATS-6* was particularly noteworthy. Capable of broadcasting to small and inexpensive ground stations, it provided experimental public health programming to remote areas in the United States and India.

# Meteorological and Imaging Satellites

The meteorological satellite program is NASA's other great applications success. The agency launched 10 TIROS (Television and Infrared Observation Satellites) into orbit between 1960 and 1965; these satellites returned a total of more than half a million high-resolution images of the earth's cloud cover, providing meteorologists with an entirely new tool for understanding and predicting the weather.

The final 2 satellites in the series, *TIROS 9* and *TIROS 10*, tested experiments and equipment that would be flown on a new generation of Tiros Operational Satellites (TOS). Once in orbit, the TOS vehicles would be called ESSA, in honor of the Environmental Sciences Services Administration, the federal agency that funded their development and would operate them in space. *ITOS 1* (Improved Tiros Operational Satellite), the first in yet another generation of meteorological spacecraft, was launched on December 11, 1970. These satellites were operated by a new federal agency in the Commerce Department, the National Oceanic and Atmospheric Administration (NOAA), and were known as NOAA satellites following launch.

The ESSA and the NOAA managed the operational weather satellite program, but NASA continued to build and launch experimental meteorological satellites. Seven Nimbus spacecraft launched between 1964 and 1978 returned the first images of the earth at night and tested a variety of other instruments that opened the way to 24-hour satellite weather coverage. The Synchronous Meteorological Satellites launched in 1974 and 1975 were transferred to the NOAA and led to the series of eight Geostationary Operational Environmental Satellite (GOES) vehicles built and launched by NASA with funding from the NOAA since 1975.

The ability to take high-resolution images of the earth's surface in a variety of spectral bands beyond the range of unaided human vision has proven valuable in a great many areas other than weather forecasting. Since 1972, the remote sensing devices flown aboard NASA's Landsat earth resources observation satellites have returned millions of images of enormous value to geologists, foresters, cartographers, agricultural planners, land and water resource managers, and city planners. Landsat data has been used to assist in the search for new oil and mineral resources; to map the spread of pollution; to predict crop production; and to assess the damage caused by earthquakes, forest fires, and floods.

Landsats 1, 2, and 3, launched between July 1972 and March 1978, demonstrated the role that satellite imagery could play in the solution of earth-bound problems. *Landsat 4*, orbited in July 1982, was turned over to the NOAA as the first operational vehicle in the system. *Landsat 5*, launched on May 1, 1984, was the first to be fully funded by the NOAA.

*A 1976 Landsat image—a mosaic of several multispectral scanner scenes—shows most of Florida and part of Georgia. Landsat satellites, formerly called Earth Resources Technology Satellites (ERTS), photograph the Earth's surface in shades of green and red and distinguish by color such features as soil, vegetation, and water; for example, green vegetation is visible in shades of red or pink, and bodies of water in blue or black. Landsat images enable scientists to forecast crop production around the world, locate energy and mineral resources, and assess population densities.*

# Exploring Space

In addition to its pioneering efforts in the field of earth applications, NASA has always used satellite technology to explore the Solar System and beyond. The first goal of the NASA space science program was to achieve a better understanding of the complex interplanetary environment in the neighborhood of Earth. Between 1958 and 1963, a series of Explorer scientific satellites and Pioneer and Mariner probes measured and mapped the flow of high-energy particles streaming out from the sun. Since the late 19th century, scientists had theorized that Earth was constantly bombarded by such particles; some believed that there was a link between this type of solar activity and Earth's weather. Others noticed a relationship between an increase in solar activity, evidenced by solar flares and sunspots, and the intensity of the northern lights (aurora borealis) and disturbances in radio communications being bounced off ionized layers in the upper atmosphere.

The first satellites and space probes unlocked the mystery. They confirmed the existence of what became known as the solar wind, a flow of charged electronic particles, electrons, and protons streaming out from the sun in all directions. The spacecraft proved that the stream of particles moves at various speeds, depending on the level of solar activity, that it conducts electricity, and that it carries with it a trapped magnetic field. When the solar wind encounters the magnetic field surrounding a planet like Earth, it is deflected and flows around the field like a stream of water around a partially submerged rock. The complex area of interaction between the solar wind and Earth's magnetic field, from the "bow shock wave" closest to the sun to the tip of the tail extending beyond the orbit of the moon, is known as the magnetosphere.

But, as the satellites revealed, not all of the solar wind is deflected around Earth. Some particles are trapped and held in two radiation belts by Earth's magnetic field. Named after their discoverer, James Van Allen of the University of Iowa, the inner radiation belt, composed of protons, is centered some 2,500 miles above the equator. The outer electron belt is located 10,000 miles above the surface. The study of the solar wind, magnetosphere, radiation belts, and other aspects of the relationship between high-energy particles and magnetic fields in space continues to the present day.

With a record of early successes behind them, in the early 1960s NASA space planners began launching a second generation of larger and more complex scientific satellites. Known as observatory spacecraft, they weighed 5 to 10 times as much as the original Explorers and could be controlled in orbit and maneuvered to a degree impossible with their predecessors.

Between 1962 and 1975, the agency successfully launched eight Orbiting Solar Observatories, six Orbiting Geophysical Observatories, and three Orbiting Astronomical Observatories (but only two actually functioned). Each series of observatory consisted of a basic spacecraft design that could carry up to 20 experiments per flight. In addition, NASA flew a series of specialized scientific spacecraft, including three biosatellites, designed to study the impact of space conditions on living organisms, and Pegasus, which gauged the amount of micrometeoroid activity in near-earth space.

The new generation of scientific spacecraft produced a wealth of information and provided astronomers, geophysicists, and astrophysicists with a revolutionary new set of tools. The second Orbiting Astronomical Observatory, launched on December 7, 1968, took the first ultraviolet photographs of stars and provided hard evidence for the existence of the incredibly dense collapsed stars known as black holes. The *Uhuru* satellite of 1970 searched the sky for X-ray sources and identified three new pulsar stars. *Orbiting Solar Observatory 7* caught the beginning of a solar flare on film for the first time and discovered "polar caps" on the sun.

NASA's program of interplanetary exploration was also flourishing. Early attention focused on Earth's closest neighbors, Venus and Mars. On December 14, 1962, the *Mariner 2* spacecraft flew past Venus at a distance of 21,625 miles. Although not equipped with cameras, the vehicle discovered that, unlike Earth, Venus has only a very weak magnetic field and no radiation belts. In addition, the first successful U.S. interplanetary explorer determined that the surface temperature of the planet was a very high 800 degrees Fahrenheit.

The next American visitor to Venus, *Mariner 5*, flew past the planet on October 19, 1967, one day after the Soviet *Venera 4* probe had achieved the first landing on the surface. Between them, the 2 spacecraft indicated that Venus has an atmosphere composed of roughly 90 percent carbon dioxide, an even higher surface temperature (980 degrees Fahrenheit) than indicated by *Mariner 2*, and a surface atmospheric pressure perhaps 100 times higher than that of Earth.

A third U.S. spacecraft, *Mariner 10*, flew past Venus in February 1974, returning detailed images of the dense clouds shrouding the surface of the inhospitable planet. After swinging past Venus, *Mariner 10* continued on to Mercury. The spacecraft completed three passes over the surface of the desolate, innermost planet in March and September 1974 and March 1975, providing the first close-up look at the face of Mercury.

NASA's most ambitious attempt to probe the mysteries of Venus came with the launch of *Pioneer Venus 1* and *Pioneer Venus 2*. The first spacecraft went

*On February 6, 1974, Mariner 10, an interplanetary satellite, sent this image of the clouds of Venus to the Jet Propulsion Laboratory. After it flew by Venus, Mariner 10 provided the first close-up pictures of the planet Mercury.*

into orbit around Venus on December 4, 1978, and would continue to return data and images of Venus for the next decade. *Pioneer Venus 2*, which arrived at its destination 5 days later, on December 9, 1978, consisted of a large main structure, or bus, 1 large probe, and 3 identical small probes. Plunging toward the planet, the spacecraft returned information on the upper atmosphere, discharged its probes, and was destroyed by the heat of entry. The four probes transmitted data during their descent to the surface. Although not designed to survive impact, 1 of the probes continued transmitting for 67 minutes.

Of all the planets, Mars seems most intriguing. It is not surprising that NASA chose to concentrate on exploring Earth's mysterious red neighbor. *Mariner 4* became the first spacecraft to conduct a flyby of Mars in July 1965, followed by *Mariner 6* and *Mariner 7* in July and August 1969. The narrow strips of surface photographs transmitted back to earth by the three spacecraft were somewhat disappointing to those who hoped for signs of a vanished civilization. The planet looked bleak and as pockmarked with craters as the surface of the moon. Like Venus, Mars lacked a magnetic field and radiation belts, but it did have a thin atmosphere.

While the Soviets concentrated on Venus, NASA returned to Mars with *Mariner 9* in November 1971. As luck would have it, the spacecraft arrived just as a gigantic dust storm had begun to sweep across the planet. When the surface of Mars began to emerge two months later, however, scientists were delighted to find a very different world from that which they had been led to expect by the flyby spacecraft. One of the first features to appear was an enormous volcanic crater, Olympus Mons, 18 miles high and 300 miles in diameter.

Far from being dead and barren, Mars had once been the scene of volcanic activity on a scale so enormous that this one crater dwarfed anything to be found on Earth. The surface was broken with fracture patterns indicating massive geological activity. There were huge canyon systems stretching across an area equal to the width of the United States and what appeared to be dried-up riverbeds. No signs of life were visible from orbit, but Mars had become a very interesting place once again.

*In July 1976,* Viking Lander 1 *photographs the terrain of Mars while the lander's surface scoop (left foreground) collects samples of the soil. NASA's Viking landers mapped the planet, analyzed the Martian atmosphere, and successfully demonstrated the use of robot spacecraft in planetary exploration.*

The Viking landings of 1976 were not only the high point of NASA's Mars exploration effort; they were in many ways the most ambitious of all interplanetary missions undertaken to date. A total of four spacecraft—two Viking orbiters and two landers—would make the journey to Mars. Launched a few weeks apart, on August 25 and September 9, 1975, they flew to Mars and went into orbit, conducting a survey of the preselected landing spots to be sure that they were suitable.

*Viking Lander 1* touched down in a basin known as the Chryse Planitia on July 20, 1976. The second lander descended to the surface of the Utopia

*Jupiter's Red Spot (left) and a shadow of the moon, Io (right), are shown in a photograph taken by* Pioneer 10 *on December 1, 1973. NASA designed* Pioneer 10 *and* Pioneer 11 *to conduct flyby surveys of Jupiter and Saturn. In 1986,* Pioneer 10 *became the first human-made object to escape the Solar System.*

Voyager 1 *photographs Saturn, its rings, and three of its satellites on August 24, 1980. NASA's unmanned Voyager missions to study Jupiter, Saturn, Uranus, and Neptune over a period of 12 years were stunningly successful; the total cost of the Voyager program was less than NASA pays annually for its Space Shuttle program.*

Planitia on September 3. For the next year the two spacecraft photographed every aspect of the surface within their view; kept tabs on local weather conditions; analyzed the composition of the atmosphere; and dropped scoopfuls of the local soil into an on-board "chemical laboratory" that reported its constituent elements and searched for traces of organic material. At the same time, the orbiters were mapping the planet and its moons and relaying information and instructions between scientists on earth and the two landers.

The Viking project did not discover life on Mars, but it did demonstrate the extent to which sophisticated robot spacecraft could thoroughly explore another world. In fact, the Viking landers kept sending data back to earth into the 1980s, long beyond their designed lifetime.

The outer planets were the final target for NASA's automated explorers. *Pioneer 10* and *Pioneer 11*, designed to conduct flyby surveys of Jupiter and Saturn, were launched on March 3, 1972, and April 6, 1973. More than a year and a half after leaving Earth, *Pioneer 10* swept through its closest approach to Jupiter on December 3, 1973. Approaching its rendezvous with the planet, the spacecraft had mapped and measured Jupiter's turbulent magnetosphere and radiation belts, measured temperatures, photographed and studied its moons, and sent back breathtaking images of the planet itself, swathed in multicolored bands of cloud. *Pioneer 11* made an even closer approach to Jupiter just a year later, on December 2, 1974, performing the same sort of survey conducted by its predecessor before swinging off toward a rendezvous with the ringed planet Saturn and its moon Titan in September 1979.

*Voyager 1* and *Voyager 2*, a pair of even more sophisticated spacecraft, followed the path to the outer planets blazed by the 2 Pioneers. Launched on September 5, 1977, *Voyager 1* flew a swifter trajectory and brushed past Jupiter in March 1979, 4 months before its companion, *Voyager 2*, which had been launched on August 20, 1977. Together, the two interplanetary travelers made new discoveries relating to the planet's complex magnetosphere and took detailed and stunning close-up photos of the planet and its four major moons.

Arriving at Saturn in the fall of 1980 and the summer of 1981, the Voyagers studied the atmosphere of the planet, photographed its cloudy surface, investigated the famous rings, and raised the number of known Saturnian moons to 15. *Voyager 1*, its mission complete, continued on toward an eventual exit from the Solar System. Like its companion, the spacecraft carried a gold-plated copper record, "The Sounds of Earth," containing two hours of sound and music, along with digitized images of Earth. The plaques aboard the Pioneer spacecraft were designed to show another civilization what humans look like and where they live. The Voyager records would provide a great deal more information.

In January 1986, *Voyager 2* became the first spacecraft to visit Uranus. Based in part on Voyager data, planetary scientists believe that the planet consists of a dense molten core about the size of Earth lying at the bottom of an ocean of water and ammonia some 6,000 miles deep. The atmosphere consists primarily of hydrogen and helium, with small amounts of methane, ammonia, and water vapor. Voyager discovered 10 new moons orbiting Uranus

*The trajectory of the two Voyager spacecraft. The spacecraft are powered by thermoelectric generators and use a large antenna dish to transmit and receive information. Each spacecraft also carries scientific equipment and two television cameras.*

and raised the number of rings surrounding the planet to 11. In August 1989, *Voyager 2* flew to within 800 miles of Neptune, then followed its companion out of the Solar System.

The enormous successes that have been achieved with automated scientific spacecraft overshadow a fundamental problem within NASA. Since the early 1960s, the scientists and engineers involved in the effort have often felt like the second-class citizens of an agency in which space science is by no means the highest priority.

The problem is reflected in the organizational structure of the agency. Under T. Keith Glennan all operational space activity, manned and unmanned, was the business of the Office of Space Flight Programs. Soon after taking over the agency, James Webb created a new Office of Manned Space Flight (later renamed the Office of Space Transportation Systems, now the Office of Space

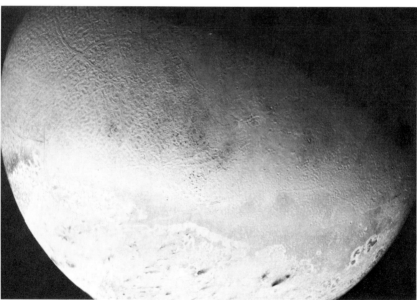

*In August 1989,* Voyager 2 *flew past Neptune (top) and Neptune's moon, Triton (bottom). Fourteen separate images from* Voyager 2 *were combined by scientists at the Jet Propulsion Laboratory to produce this view of Triton.*

Flight) to manage astronaut activity. Satellites and space probes became the responsibility of an Office of Space Science and Technology. With the segregation of the total space effort, the budget for the popular and highly visible manned programs grew enormously, whereas the proportion of total funds allocated to unmanned science and applications projects was drastically reduced.

It was the beginning of a serious rift that continues within the agency to the present day. The lion's share of the funding has gone to support the enormously expensive and dramatic manned programs—Mercury, Gemini, Apollo, and the space shuttle. Important scientific missions, such as the exploration of Halley's comet at the time of its return to earth in 1985–86, and the Galileo program, designed to send both an orbiter and a lander to Jupiter in 1986, were postponed or abandoned for lack of funding.

The advent of the space shuttle, a space truck specifically designed to carry satellite payloads into orbit, promised to draw the two segments of the total U.S. program back together. Instead, launch delays and a total halt to shuttle flights following the *Challenger* disaster have kept revolutionary new scientific spacecraft like the Hubble Space Telescope on the ground.

It is important to remember, however, that although space science may have suffered hard times within NASA during the post-Apollo years, it has by no means come to a halt. Since 1977, the agency has continued to build and fly spacecraft that look at the heavens in new ways—three High Energy Astronomy Observatories; the International Ultraviolet Explorer; an Infrared Astronomy Satellite; Helios, which carried instruments closer to the sun than ever before; the International Sun-Earth Explorer; and the International Radiation Investigation Satellite, to name but a few. With the return of the space shuttle to operational status, there is every reason to hope that the long-sought balance between manned and unmanned programs will be achieved at last.

*On its maiden voyage, the space shuttle* Columbia *rises off the launchpad at Kennedy Space Center on April 12, 1981. NASA designed the shuttle to be totally reusable except for its external fuel tank; after its spaceflight the craft will glide to an airplanelike landing at Edwards Air Force Base in California's Mojave Desert.*

# EIGHT

# The Shuttle Era

The conquest of space sounded so simple when Wernher von Braun described it in the 1950s. It would be a step-by-step process, beginning with the launch of unmanned satellites, followed by short trips into space aboard winged rockets. Eventually, those reusable winged spacecraft would begin shuttling building materials into space so that work could begin on a permanent earth-orbiting space station that would serve as a base for later manned voyages to the moon, to Mars, and beyond.

NASA leapfrogged that systematic plan, flying to the moon without developing either a reusable winged shuttle craft or a manned space station. Mercury, Gemini, and Apollo achieved the goal of a lunar landing, but they did not provide the foundation for a long-term future in space. Even before the first step on the moon, NASA scientists and engineers realized that any post-Apollo planning would have to include both a winged spacecraft and an orbital base: These projects eventually took the form of the space shuttle and Skylab.

## Skylab

The experimental space station known as Skylab grew out of a series of discussions between officials at NASA headquarters and engineers at Huntsville and Houston. Anxious to develop a relatively inexpensive but useful and

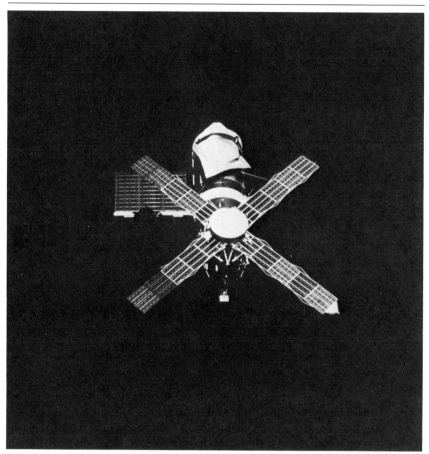

*An overhead view of the Skylab space station, taken during the final "flying around" inspection in June 1973. NASA's earth-orbiting space station was designed to prove that men could work and live in space for long periods of time without ill effects.*

impressive follow-on program to the moon landings, planners came up with the notion of flying a manned orbiting station using spare Apollo hardware.

The final plan called for a crew of three astronauts to take up residence inside the specially modified third (S-4B) stage of a Saturn 5 rocket. Fifty-nine feet long and almost 22 feet in diameter, the interior of the S-4B would be transformed into a comfortable home complete with 3 small bedrooms; a kitchen-dining area; a bathroom with a toilet and shower designed for use in zero gravity; an exercise area and specialized work spaces. Skylab would also be equipped with a medical facility, photo lab, broadcasting studio, and a sophisticated astronomical observatory. Enormous solar panels attached to the exterior of the S-4B would generate more than enough electricity to power the station.

The refurbished S-4B would be launched into a stable earth orbit as the unpowered third stage of the Saturn 5. The crew, launched separately aboard a smaller Saturn 1-B, would dock their standard Apollo CSM to a special module attached to the upper end of the Skylab, enter the station, and settle down for a long-term stay in space. Unlike Apollo, where there had been little time for real science, Skylab would enable astronauts to conduct a wide variety of experiments in areas ranging from astrophysics to the production of new materials in space. There would be opportunities to investigate the medical consequences of long stays in a weightless environment and time to conduct a full program of earth observation photography.

The last Saturn 5 ever to fly carried Skylab into orbit on May 14, 1973. The program almost came to an end before it began when excessive vibration

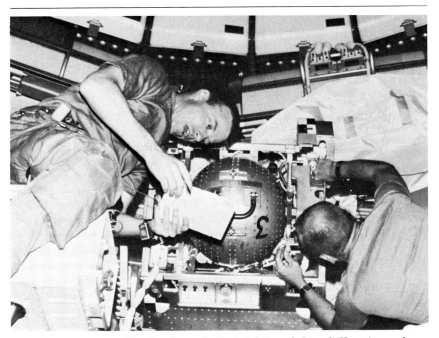

*Skylab 2 astronauts Charles Conrad, Jr. (right) and Joseph Kerwin work on an experiment aboard Skylab in June 1973. The astronauts spent 28 days in Skylab conducting scientific and medical experiments.*

ripped a protective micrometeoroid and sun shield and one of the two huge solar arrays away from the S-4B stage during the ascent. Unable to deploy the second large solar array, which was held in place by a piece of the broken shield, ground controllers maneuvered a smaller solar panel designed to power scientific instruments into position facing the sun. Telemetry reported that the spacecraft was now receiving some power but that the temperature inside the cabin was rising to dangerous levels without the protection of the shield.

The first Skylab crew, Charles Conrad, Jr., Paul Weitz, and Joseph Kerwin, were launched on May 25, 1973. Uncertain what the astronauts would find when they reached the disabled spacecraft, NASA engineers equipped them with an assortment of tools and a hastily devised substitute sunshade. Over the next two days, operating out of the cramped Apollo CSM and the sweltering interior of Skylab, the crew succeeded in deploying the new shield and freeing the undamaged solar panel. Skylab was saved.

Three astronaut crews—9 men—would live and work in Skylab for a total of 171 days and 13 hours between May 25, 1973 and February 8, 1974. During that time they conducted nearly 3,000 scientific, medical, and engineering experiments. They overcame problems as well. The astronauts found that it took time to become accustomed to weightlessness. They suffered bouts of illness in orbit, complained about bland food, and voiced resentment over the full work schedules prepared by mission planners. Complaints about overwork from the third crew led to a more reasonable schedule, complete with additional time for exercise and even a bit of relaxation.

In spite of the difficulties, the Skylab program represented NASA at its best. At minimum expense, the agency was able to operate an entirely new and extraordinarily useful project that returned information of enormous value to science, medicine, and industry. To a much greater extent than Apollo, Skylab demonstrated just how useful human beings could be in space.

The three manned Skylab missions brought the program to a close. Congress refused to fund a proposal to launch a second Skylab, and the original spacecraft was doomed. Even at an altitude of 275 miles, Skylab was fighting a losing battle with air resistance. Traces of the atmosphere gradually slowed its speed, dropping it ever closer to inevitable destruction in a fiery reentry. The end came on July 11, 1979. After 6 years and 34,981 orbits of the earth, Skylab plunged back into the atmosphere and disintegrated. Most of the chunks of the spacecraft that survived the long fall back to earth fell into the Indian Ocean. A few small bits and pieces were later found in the outback of Western Australia.

## Skylab Missions

| | | | |
|---|---|---|---|
| Skylab 1 | May 14, 1973 | Launch of the unmanned spacecraft | |
| Skylab 2 | May 25–June 23, 1973 | Charles Conrad, Jr.; Paul J. Weitz; Joseph P. Kerwin | 28 days, 50 minutes |
| Skylab 3 | July 28–September 25, 1973 | Alan J. Bean; Jack R. Lousma; Owen K. Garriott | 59 days, 11 hours |
| Skylab 4 | November 16, 1973–February 8, 1974 | Gerald P. Carr; William R. Pogue; Edward G. Gibson | 84 days, 1 hour |

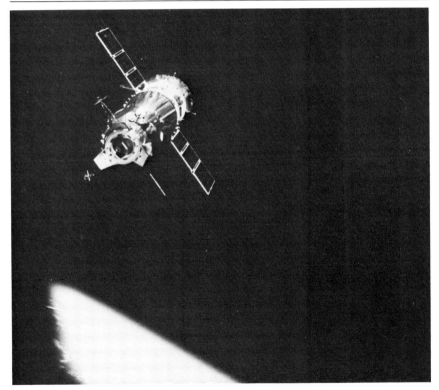

*The Soyuz spacecraft is photographed by the Apollo crew during the Apollo-Soyuz Test Project in July 1975. NASA astronauts and Soviet cosmonauts rendezvoused in space, docked their crafts together, and paved the way for future joint missions or space rescue efforts.*

NASA officials had yet another notion for putting surplus Apollo hardware to work. The Apollo-Soyuz Test Project (ASTP) was the result of talks between President Richard M. Nixon and Soviet premier Aleksey Kosygin in 1972. In view of the important role that space had played as a competitive arena during the 1960s, the two world leaders decided to underscore the new spirit of détente, or relaxation of tensions, between their nations by undertaking a joint space mission.

*Astronaut Thomas P. Stafford (left) and cosmonaut Aleksei A. Leonov meet in the hatchway of Apollo's docking module during the Apollo-Soyuz Test Project. During the two days they were linked together, the crews visited each other, conducted joint experiments, and even shared meals.*

The basic idea was simple enough. An Apollo CSM and a two-man Soyuz spacecraft would rendezvous and dock in orbit. It would not be easy. The American and Soviet spacecraft featured incompatible docking mechanisms and life-support systems that operated with different atmospheric gases at different pressures. Those differences were the best of all possible reasons for flying the mission, however. The ASTP would not only demonstrate the determination of the United States and the Soviet Union to cooperate, it would also bring American engineers and flight crews together with their Soviet counterparts, promote technical exchange, and prepare the way for future joint missions or international space rescue efforts.

Slowly but surely, over a period of two years, the U.S. and Soviet teams worked out the difficulties. ASTP astronauts Thomas P. Stafford, Vance D. Brand, and Donald K. "Deke" Slayton launched from the Kennedy Space Center on July 15, 1975, seven hours after cosmonauts Aleksei Leonov and Valery Kubasov flew into orbit from a Soviet launch complex. Forty-five hours later U.S. mission commander Stafford nosed the new docking adapter up to the Soyuz spacecraft and locked it in place. The five astronauts greeted one another in Russian and English, exchanged keepsakes, and began two days of joint scientific experiments. The Soyuz crew returned to earth 43 hours after separation. The Apollo astronauts brought the nearly flawless mission to a close with a safe splashdown on July 24.

# The Space Shuttle

Skylab and the ASTP brought the Apollo era to a close. The first model of an entirely new generation of spacecraft, the space shuttle, was rolled out of the final assembly building of the North American Rockwell facility at Palmdale, California, in September 1976. The story of the space shuttle had begun seven years earlier, however, on September 15, 1969, when a special Space Task Group headed by Vice-president Spiro Agnew presented President Richard Nixon with three options for the future of the U.S. space effort.

The Space Task Group reported that, for $10 billion a year over the next 6 years, NASA could send human beings to Mars, build a lunar base, establish an earth-orbiting space station, and develop a winged, reusable spacecraft that would shuttle back and forth between Earth and the space station. The task group's second option, at about half the price, included development of both the space station and space shuttle. Finally, a $3 billion bargain-basement special would mean the end of manned space flight by 1974 but would include some funding for future development.

*A space shuttle is assembled at the North American Rockwell facility in Palmdale, California, in 1976. After the shuttle has been assembled it will be transported to NASA's Dryden Flight Test Center, where it undergoes approach and landing tests.*

President Nixon chose the third and least expensive option but left the door open for NASA to return with a proposal for additional funding to cover a no-frills manned program. Administrator Paine recognized that Apollo, for all of its success, had not provided a firm foundation for an ongoing U.S. manned space effort. With plans for the experimental Skylab space station already under way, he decided to concentrate on obtaining approval for a fully reusable space shuttle—a winged vehicle that would become the cornerstone of all future programs.

The shuttle was to be a totally reusable, all-purpose space truck. Each vehicle would make repeated trips into space, hauling any and all satellites aloft, orbiting the earth while the crew of astronauts and mission specialists conducted their experiments, and carrying aloft materials with which to build a permanent station in space. The basic vehicle would operate only in earth orbit. A special "space tug"—essentially a strap-on rocket engine—would boost lunar and interplanetary spacecraft carried into orbit by the shuttle on to their destinations.

After the enormous expenditures on Apollo, selling the shuttle to the president and Congress was a difficult task. From the outset, NASA officials had to compromise to make their case. Prior to his departure from the agency late in 1970, Thomas Paine had decided that the shuttle could be sold only as the complete answer to the nation's space requirements for the immediate future. With the shuttle, he would argue, there would be no more need for either NASA or the military services to purchase traditional "throwaway" launch vehicles.

Air force officials, aware that NASA would need their cooperation and support to obtain approval for the new spacecraft, began to insist on some measure of design control long before they agreed to book payload space aboard the proposed shuttle. Paine's successors, first Acting Administrator George Low then Administrator James C. Fletcher, accepted far more drastic compromises.

NASA officials had originally estimated that the minimum cost for the development of a fully reusable vehicle would be $10–$15 billion. Initial budget cuts forced a fundamental redesign of Paine's fully reusable "dream machine" into a partially reusable compromise. The original plan for a large manned booster stage that would be flown back to earth after starting the second-stage orbiter on its way toward space gave way to a pair of recoverable solid propellant boosters and an enormous externally mounted liquid propellant tank to fuel the three engines of a smaller orbiter. The external tank would be jettisoned in space and destroyed during reentry.

The redesigned shuttle would be smaller and more expensive to operate than its larger, fully reusable predecessor, but it could be developed at roughly half the cost. By the early 1970s, however, officials in the Bureau of the Budget were insisting that still further cost reductions would be necessary. Representative Olin Teague of Texas and other supporters of NASA in Congress fought for the shuttle program, but allocations continued to fall below requests.

NASA officials believed that they had little choice but to forge ahead with the money allocated. The agency had bet its entire future on the shuttle. With the completion of the Skylab and ASTP programs, no more Americans would fly into space until the shuttle was operational. Satellite and planetary exploration programs were dependent on the new vehicle as well. There were no funds available for the development of more traditional launchers, and the existing stock of such rockets was dwindling.

As they struggled to overcome serious technical problems with the shuttle's main engines, the ceramic tiles designed to protect the spacecraft during reentry, and other key elements of the total system, NASA officials were convinced that they were "making do" in the best tradition of the agency. As had been the case before the Apollo 204 fire, however, a handful of critics attempted to call attention to what they regarded as potentially dangerous flaws overlooked in the rush to build and fly the new spacecraft with constantly shrinking resources.

*The space shuttle* Enterprise *is carried piggyback by a Boeing 747 during flight tests at Edwards Air Force Base in 1977. After the shuttle separated from the 747, it landed safely in the Mojave Desert five and a half minutes later.*

*In November 1982, the crew of the space shuttle* Columbia *deploy a communications satellite from the cargo bay—the first time ever an earth-orbiting
satellite was released from a manned spacecraft.*

By January 1977, it looked as though NASA decision makers had won their
bet on the shuttle. The first vehicle, Orbiter 101, named the *Enterprise* in
honor of the starship featured on the television series "Star Trek," was loaded
on a 90-wheel trailer at the North American Rockwell plant in Palmdale,
California, and trucked a short distance to NASA's Dryden Flight Test Center.
In time, there would be four more Shuttle Orbiters: *Columbia, Challenger,
Atlantis,* and *Discovery.*

When operational, the shuttles would be launched from the Kennedy Space Center in Florida and land at Edwards Air Force Base, California, adjacent to the NASA Dryden facility. The Orbiter would then be hoisted into place on top of a specially equipped Boeing 747, officially known as the Shuttle Carrier Aircraft (SCA), and flown back to Cape Canaveral for its next voyage into space. The first series of five airborne test flights of the new vehicle were conducted with the *Enterprise* riding piggyback on the SCA. On August 12, 1977, the crew broke away from the airborne 747 and flew the unpowered *Enterprise* back to a safe landing at Edwards. Four more glides of this sort completed the Orbiter test program.

*Columbia*, the first shuttle to fly into orbit, roared aloft from Cape Canaveral on August 12, 1981, with astronauts John W. Young and Robert L. Crippen at the controls. The four Orbiters made 25 trips into space between that date and January 28, 1986, when the *Challenger* lifted off on a mission officially known as STS 51-L.

The program had accomplished a great many of the things that NASA planners had expected of it. Americans were flying into space once again. One hundred and fifteen men and women took the ride of their life into space aboard the shuttle. They spent a total of 61 days, 2 hours, and 18 minutes in orbit, conducting experiments, deploying new satellites, repairing old ones, and returning still others to earth for refurbishing. One senior astronaut, Crippen, made four flights. Fifteen shuttle crew members, including one woman, Kathryn Sullivan, walked in space.

In spite of the operational successes, however, the shuttle program was not living up to the promises made for it. The enormous savings in launch costs predicted by shuttle planners were never achieved. The estimates had been based on a full payload bay for every mission and a heavy schedule of flights. In fact, NASA was forced to subsidize launch costs to keep its payload bays even partially full. To cover the cost of shuttle development and operation with launch fees, NASA would have had to charge 10 times its standard rate for sending payloads into orbit. At that price, the Orbiters would have flown with virtually empty payload bays. Commercial satellite companies, industrial firms interested in space manufacturing experiments, and other commercial users would have taken their business to the European Space Agency or to other nations with a launch capability.

Nor was NASA able to maintain its predicted launch rate. Original estimates called for a turnaround time of two weeks between missions. In practice, NASA engineers and technicians were forced to work overtime to reduce the interval between flights to two months. The Orbiters required more extensive refurbishing following each flight than had been predicted. The availability of spare parts was a constant problem.

The agency also faced continuing budget problems. NASA officials worked hard to convince the public and Congress of the importance of their program. The agency sent two of its congressional friends, Senator Jake Garn of Utah and Representative C. William Nelson of Florida, into space aboard the shuttle. Senator Garn flew as a payload specialist on mission STS 51-D, April 12–19, 1985. Congressman Nelson followed as a member of the crew of STS 61-C, January 12–18, 1986.

*Senator Jake Garn of Utah (center) experiences weightlessness in preparation for his flight on the space shuttle* Discovery *in 1985. NASA sent Senator Garn into space as a payload specialist to help deploy a satellite and to assist in medical experiments.*

113

NASA worked hard to keep its programs in the public eye. The agency publicized inexpensive "getaway special" launch rates for small experimental packages in order to drum up additional business and focus attention on the program. The agency held a nationwide student science competition and flew the winning experiments into space at no charge. An announcement that NASA would fly a teacher aboard the shuttle drew applications from 11,000 educators across the nation.

The winner of the Teacher in Space competition, Christa McAuliffe, from Concord, New Hampshire, was scheduled to fly aboard the *Challenger* on mission STS 51-L. Francis Richard Scobee, a veteran of one previous flight, was the mission commander. Rookie Michael Smith flew as pilot. Mission specialists Gregory Jarvis, Ronald McNair, Colonel Ellison Onizuka, and Judith Resnick would be responsible for the scientific and experimental work conducted in orbit.

*The last crew of the space shuttle* Challenger *included (left to right, front) astronauts Michael Smith, Francis Richard Scobee, and Ronald McNair; (back) astronaut Ellison Onizuka, payload specialist and Teacher in Space Project winner, Christa McAuliffe, payload specialist Gregory Jarvis, and astronaut Judith Resnick.*

*The space shuttle* Challenger *explodes on January 28, 1986, killing all seven crew members.* Challenger *was NASA's veteran orbiter and had carried a total of 53 crew members into space before the fatal accident.*

The payload for the planned six-day flight included a NASA communications satellite, TDRS-B/IUS, and a small Spartan spacecraft developed by NASA and the University of Colorado to study Halley's comet. There were also plans to monitor the comet from the Orbiter throughout the mission. In addition, the mission specialists would be conducting four major experiments, including one of the winning entries in the NASA Student Involvement Program for high school science students.

*Challenger* lifted off Pad 39C late on the morning of January 28, 1986. It was NASA's veteran Orbiter, having carried a total of 53 crew members into space during the course of 9 previous missions. One minute and 13 seconds into its 10th flight, *Challenger* vanished in a cloud of smoke and flame. Gus Grissom, Ed White, and Roger Chaffee were no longer the only casualties of the U.S. space effort.

*An artist's rendering of* Freedom, *the U.S. space station under development and scheduled for permanent occupancy in orbit by the mid-1990s. Public reaction to the project has been mixed—taxpayers question* Freedom's *cost, and some scientists question NASA's priority of manned rather than un-manned space programs.*

# NINE

# Toward the Future

On September 29, 1988, 32 months after the *Challenger* tragedy, the space shuttle *Discovery* lifted off from Cape Canaveral. For the men and women of NASA, it had been a difficult two and a half years. In the immediate aftermath of the tragedy, President Ronald Reagan appointed William P. Rogers, former attorney general and secretary of state, to head a distinguished commission charged with investigating the accident and recommending a course of action for the future. The members of the Rogers Commission worked quickly, delivering their final report in only four months.

The commission had little difficulty identifying the specific cause of the disaster. The rubber O-rings sealing the joint between the bottom two segments of the right solid propellant booster had failed, allowing flames to burn through the outer casing of the booster and then through the skin of the external liquid propellant tank.

But the problems went much deeper than the failure of a single part of the spacecraft. Engineers at Morton Thiokol, Inc., the firm that built the solid boosters, had expressed serious concern about the performance of the pressure seals. A study of seals recovered from previous flights had revealed several instances of damaged O-rings. The worst examples came from launch days when it had been cold at the Kennedy Space Center.

On the day before the *Challenger* flight, as it became apparent that this would be the first shuttle launch conducted with the temperature at Cape

Canaveral below freezing, a series of telephone conferences took place between nervous Thiokol engineers in Utah and NASA solid booster specialists at Huntsville. At one point, a Thiokol expert recommended that the *Challenger* launch be postponed until the temperature reached 53 degrees, the previous record low for a launch. When NASA officials questioned the necessity of such a step, Thiokol managers withdrew their objections.

At NASA, the desire to keep the program moving had overwhelmed considerations of safety. Agency officials had dismantled the intricate network of quality assurance procedures instituted by James Webb, returning primary responsibility for safety to program managers and the heads of the individual research centers. No one in authority at NASA headquarters, Houston, or Cape Canaveral had played a serious role in the afternoon and evening debate relating to the seals on the solid boosters. Officials at Huntsville, the center responsible for supervising work on the boosters, handled the matter.

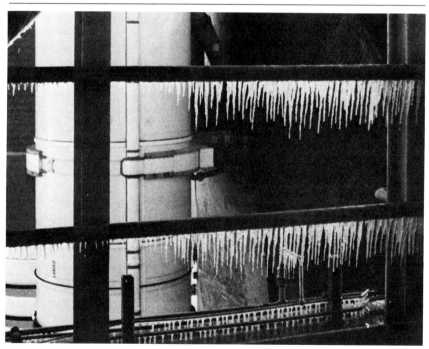

*A view of the space shuttle* Challenger *on the launchpad during an ice inspection on the morning of January 28, 1986. The shuttle launch at Cape Canaveral would be the first one conducted with the temperature below freezing.*

*William P. Rogers, chairman of the Presidential Commission on the Space Shuttle Challenger Accident, listens to testimony during a hearing. The Rogers Commission determined the reason for the shuttle's explosion: The rubber O-rings sealing the joint between the bottom two segments of the right solid propellant booster had failed and had allowed flames to burn through the booster wall and reach the external fuel tank.*

In addition to pinpointing the exact cause of the accident, the members of the Rogers Commission provided a list of other potentially catastrophic technical problems. They made nine basic recommendations, ranging from the obvious suggestion that the solid rocket booster joints be redesigned to more far-ranging proposals relating to the management of the agency.

The committee recommended that final responsibility for safety be placed in the hands of an independent group reporting directly to the administrator, as had been the case under James Webb. In addition, they suggested that NASA managers improve communications between the centers and headquarters; undertake an immediate and thorough safety review of all shuttle systems; address the problem of landing safety; and develop an improved crew escape system. Finally, the members of the Rogers Commission advised that NASA establish a slower and more reasonable launch schedule and take whatever steps were necessary to ease the "relentless pressure" of operating the nation's only space vehicle.

The report was blunt and honest, identifying major problems that had been growing throughout the post-Apollo years. At the same time, the members of

*In April 1986, NASA scientists inspect segments of the solid rocket booster slated for a future shuttle mission in order to show investigators of the* Challenger *accident the way the segments seal once they are in place for a launch.*

the commission were careful to express their confidence in the ability of the agency to correct its problems and urged "that NASA continue to receive the support of the Administration and the nation." They added:

> The agency constitutes a national resource that plays a critical role in space exploration and development. It also provides a symbol of national pride and technological leadership. The Commission applauds NASA's spectacular achievements of the past and anticipates impressive achievements to come. The findings and recommendations presented in this report are intended to contribute to the future NASA successes that the nation both expects and requires as the 21st century approaches.

The Apollo 204 fire occurred at a time when NASA was racing to achieve a monumental goal. Morale within the agency was at its peak. The need to reorganize and push forward to accomplish the primary objective of a lunar landing helped to pull NASA through the crisis. The *Challenger* disaster struck an agency that was already hard-pressed from attempting to meet impossible schedules and deadlines.

Morale plummeted in the wake of the tragedy. James Beggs, the administrator of NASA, had resigned two months before the *Challenger* crash. Now other top officials either resigned or were reassigned. Center personnel at Huntsville, Houston, and Cape Canaveral squabbled over responsibility for the O-ring problem. NASA, the agency with the reputation for moving at top speed, seemed to be running out of steam. At times, progress toward assessing and correcting the problems of the shuttle seemed very slow.

With the shuttle grounded, it was obvious that the era of the expendable launch vehicle was by no means over. The U.S. Air Force, struggling to maintain the all-important reconnaissance satellite system with a dwindling supply of aging Titan 3 and Delta launch vehicles, had little difficulty persuading Congress to approve the use of the new Titan 4 as a satellite launcher. COMSAT/Intelsat and other commercial satellite firms began to explore the possibility of flying aboard Soviet, European, Japanese, or Chinese launchers. The Reagan administration, anxious to encourage the growth of a U.S. commercial launch industry, announced that in the future the shuttle would fly only those payloads specifically designed for it or which are vital to the defense of the nation.

Today, NASA is still recovering from the *Challenger* disaster. A much improved space shuttle is back in business, operating on a more reasonable launch schedule. James C. Fletcher, who served as administrator from 1970 to 1977, took over the post again in 1986 and appointed both a new deputy administrator and a new assistant administrator for Space Flight, Richard Truly, a veteran shuttle astronaut. With the agency moving forward once again, Fletcher handed over the reins to Truly, who in 1989 became the eighth NASA administrator.

Fletcher and Truly have taken steps to revitalize the lagging planetary exploration program. A sophisticated Venus probe, *Magellan*, was launched in April 1989 by the crew of space shuttle mission STS-30. Equipped with an advanced imaging radar system capable of penetrating the thick clouds that blanket Venus, *Magellan*'s mission is to map 90 percent of the planet's surface.

What does the future hold for NASA? Some things are fairly certain. There will be a U.S. manned orbiting space station in the not too distant future. President Reagan approved the project in January 1984, and while many of the major design decisions have yet to be made, the plans are slowly falling into place.

*On September 29, 1988,* Discovery *is launched from Kennedy Space Center. During the space shuttle's four-day mission it deployed a tracking and data relay satellite for use in air-to-ground voice communications and television transmissions from future space shuttle missions.*

122

*Launched on February 20, 1986, the Soviet space station Mir (Peace) is 43 feet long, has a diameter of 13.7 feet, and weighs more than 46,300 pounds. Since 1971, the Soviets have placed seven space stations in orbit. NASA plans to launch its second space station in the mid-1990s.*

At best, reaction to the space station has been mixed. One major newspaper has referred to the project as "an expensive yawn in space." Plans for the station have also underscored some of the agency's long-standing problems. Many space scientists, for example, are convinced that the project is proof that, as in the past, manned efforts will continue to take precedence over scientific satellites and planetary programs. There has been serious talk of the need to set aside 20 percent of NASA's annual congressional appropriation for space science. The agency would be required to spend that portion of its funds as directed by a committee of scientists.

In spite of the continuing arguments over priority for manned and unmanned programs, the lack of public enthusiasm, shrinking budgets, a rather vague timetable, and an assortment of other problems, the future of the U.S. space station seems assured. NASA regards such a craft as an absolute requirement for long-term manned orbital operations, an area in which the United States has fallen behind the Soviet Union. Since 1971, a steady stream of cosmonauts have spent months at a time living and working aboard ever more sophisticated Salyut and Mir (Peace) orbiting space stations. Once again, NASA is in the position of having to catch up with the USSR.

More important, the program will provide the essential cornerstone for future U.S. manned efforts. A multipurpose facility, the station will be a center for research in a variety of space sciences; a stable orbital platform for earth observations; and an experimental laboratory in which to produce pharmaceuticals, metals, crystals, and other items under zero-gravity conditions. The station may also help to inaugurate a new era of international cooperation in space.

And what of the more distant future? Agency officials have been giving a great deal of thought to the subject. In 1987, Dr. Sally Ride, the first American woman to fly into space and a member of the Rogers Commission, headed a task force established by Administrator James Fletcher to suggest future directions for NASA. In her report, *Leadership and America's Future in Space*, Dr. Ride notes that the U.S. civilian space program stands at a crossroads:

> Two fundamental, potentially inconsistent views have emerged.
> Many people believe that NASA should adopt a major, visionary goal.
> They argue that this would galvanize support, focus NASA
> programs, and generate excitement. Many others believe that NASA
> is already overcommitted in the 1990s; they argue that the space
> agency will be struggling to operate the Space Shuttle and build the
> Space Station, and could not handle another major program.

Pointing out that the United States can no longer hope to maintain leadership in every area of spaceflight, the members of the task force identified four basic initiatives:

• Mission to Planet Earth: With the space station and nine specialized satellites in orbit by the year 2000, NASA and cooperating international partners could keep a careful watch over global cloud cover; changes in the earth's vegetation; alterations in the polar caps; worldwide rainfall and moisture; the chemical composition of the oceans; shifts in the earth's tectonic plates; and changes in the structure and composition of the atmosphere.

• Exploration of the Solar System: NASA has a radar-mapping mission to Venus under way, and the long-delayed and often-postponed Galileo spacecraft was finally launched on October 18, 1989, to begin its six-year journey to Jupiter. NASA hopes to send an Observer spacecraft to Mars during the years 1989–1992. There are no plans for additional missions beyond that point. The Soviets plan a very ambitious program for the same period, including a return mission to Mars that will retrieve samples of Martian soil and rock.

*In 1983, Sally Ride became the first American woman to fly into space. In 1987, Ride, who is a civilian and has a Ph.D. in physics, headed a task group created to evaluate the NASA space program.*

*On May 4, 1989, the* Magellan *spacecraft is released into space by the crew of the* Atlantis. Magellan's *mission is to create a detailed map of the surface of Venus. The radar mapping mission was strongly supported by the task group established by NASA Administrator James C. Fletcher in 1987 to suggest future directions for the agency.*

The task force believes that it is essential to maintain U.S. leadership in the field of automated planetary exploration. In order to accomplish that, it suggests adding three new missions to the current program. A Comet Rendezvous Asteroid Flyby (CRAF) spacecraft, to be launched in 1993, would study the asteroid Hestia and pass within 25 kilometers of the comet Tempel 2. Another robot explorer, Cassini, would reach Saturn in 2005. In addition to studying the planet from orbit for three years, Cassini would launch a series of probes into the atmosphere of Saturn and Titan, its largest moon. Ideally, the spacecraft would also dispatch a lander to Titan. Finally, three unmanned Mars exploration missions would be launched from the space station between 1996 and 1999. Each would visit a different section of the planet and be equipped with an automated roving vehicle capable of traversing the Martian surface and collecting soil samples for return to the earth-orbiting space station.

- Outpost on the Moon: The third initiative suggested by the task group calls for a sophisticated, unmanned reconnaissance of the moon beginning in the 1990s, the object of which will be to locate a site for the first permanent lunar base. The first construction crew would arrive on the moon sometime during the period 2000–2005. By the year 2010, the base would be able to support 30 human beings for months at a time. In addition to conducting research in a broad range of scientific disciplines, the lunar colonists would undertake mining studies and pursue experiments in materials processing.

- Humans to Mars: The final step in the task force's program calls for the establishment of a human base on Mars during the second decade of the 21st century. The first Martian colonists would be preceded by a new generation of robot explorers. Their spacecraft would be assembled in earth orbit, with the space station serving as a construction base. In the words of Dr. Ride's report: "A successful Mars initiative would recapture the high ground of world space leadership and would provide an exciting focus for creativity, motivation, and pride of the American people. The challenge is compelling, and it is enormous."

Dr. Ride and her colleagues have presented what they regard as a minimum program to preserve U.S. leadership in selected areas of space science, applications, and manned exploration. There is little doubt that NASA could accomplish all of those things over the next three decades, given the necessary funding. Unfortunately, there is nothing to indicate that the enormous budgets required to support a return to the moon or human voyages to Mars will be available in the foreseeable future. Perhaps the most important lesson of the *Challenger* relates to the difficulties and dangers faced in forging ahead with an ambitious but inadequately funded program.

Barring the return of another era of overwhelming political and public support for spaceflight of the sort that sent the Apollo astronauts to the moon, NASA will have to proceed slowly and cautiously. The agency will have its hands full flying the shuttle, building the space station, and maintaining its unmanned applications and space science programs for the next decade or so.

Perhaps, in the not too distant future, Americans will live in colonies on the moon and Mars. It is unlikely, however, that the government of the United States, or of any other nation, will be willing or able to bear the full cost of such ventures. International cooperation on a variety of levels is already an important feature of both manned and unmanned space operations. In 1987, Soviet premier Mikhail Gorbachev suggested that the USSR and the United States consider the possibility of future large-scale joint missions. Rather than

*On October 3, 1988, the space shuttle* Discovery *returns to Edwards Air Force Base in California after its four-day mission in space. With the success of the reusable space shuttle, NASA has preserved U.S. leadership in space transportation.*

racing America's rivals to Mars, there is every possibility of flying there together, sharing the expenses, and the glory.

Of one thing there can be no doubt. Human beings are born to explore. It is part of their very nature to wonder what lies over the next hill. Whatever the cost, they will never be able to resist the challenge of the endless frontier of space. American physicist Robert Goddard said it best in a letter written to the British novelist H. G. Wells in 1932.

> There can be no thought of finishing, for "aiming at the stars," both literally and figuratively, is a problem to occupy generations, so that no matter how much progress one makes, there is always the thrill of just beginning.

# Appendix:
# NASA Administrators

| | |
|---|---|
| T. Keith Glennan | 1958–61 |
| James Webb | 1961–68 |
| Thomas Paine | 1969–70 |
| James Fletcher | 1971–77 |
| Robert Frosch | 1977–81 |
| James Beggs | 1981–86 |
| James Fletcher | 1986–89 |
| Richard Truly | 1989– |

# National Aeronautics and Space Administration

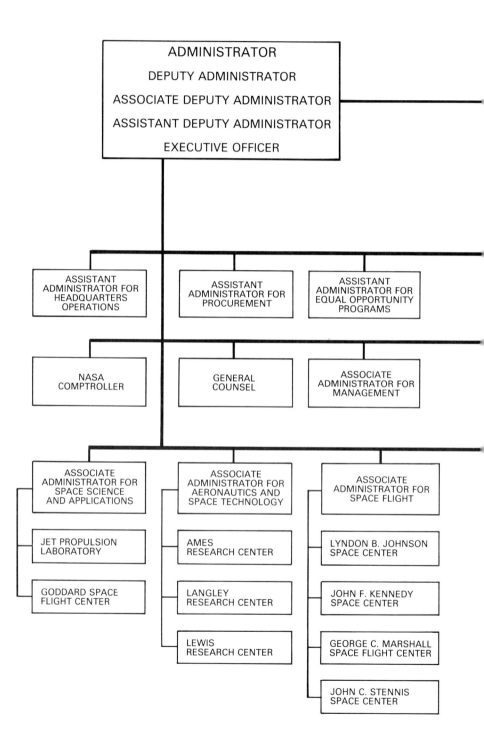

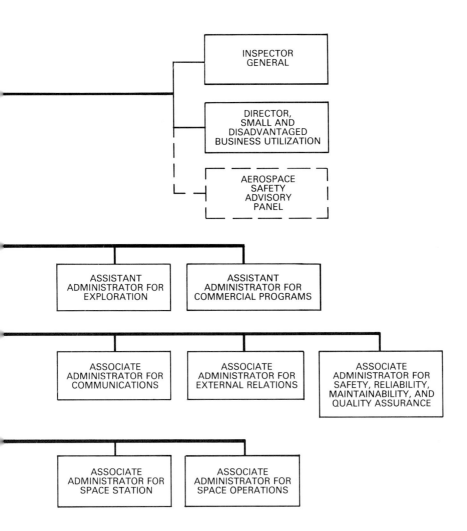

INSPECTOR
GENERAL

DIRECTOR,
SMALL AND
DISADVANTAGED
BUSINESS UTILIZATION

AEROSPACE
SAFETY
ADVISORY
PANEL

ASSISTANT
ADMINISTRATOR FOR
EXPLORATION

ASSISTANT
ADMINISTRATOR FOR
COMMERCIAL PROGRAMS

ASSOCIATE
ADMINISTRATOR FOR
COMMUNICATIONS

ASSOCIATE
ADMINISTRATOR FOR
EXTERNAL RELATIONS

ASSOCIATE
ADMINISTRATOR FOR
SAFETY, RELIABILITY,
MAINTAINABILITY, AND
QUALITY ASSURANCE

ASSOCIATE
ADMINISTRATOR FOR
SPACE STATION

ASSOCIATE
ADMINISTRATOR FOR
SPACE OPERATIONS

# GLOSSARY

**Aerodynamics**   The study of the flow of air and of the forces acting on bodies of various shapes, sizes, and characteristics moving through it.

**Booster**   An auxiliary part of the propulsion system of a rocket used to supply a part or all of the thrust during the launching and initial stage of flight.

**Command and Service Module (CSM)**   The vehicle used in the Apollo program to transport astronauts to the moon's orbit and back to Earth.

**Communications Satellite Corporation (COMSAT)**   An organization owned and controlled half by the government and half by large communications companies set up in 1962 to contract with aerospace companies for the design and construction of communications satellites and to pay NASA to launch them.

**Deep space**   The region beyond Earth's atmosphere, including space outside the Solar System.

**Geosynchronous satellite**   An artificial satellite that travels above the equator at the same speed as the rotation of Earth.

**Lunar Module (LM)**   The transport vehicle used in the Apollo program to ferry two astronauts between the Command and Service Module in lunar orbit and the surface of the moon.

**National Advisory Committee for Aeronautics (NACA)**   A federal agency founded in 1915 to conduct basic research to advance American aviation technology.

**Payload**   In unmanned spacecraft, the data-collecting and transmitting equipment. In manned spacecraft, the personnel, life-support systems, and equipment necessary to accomplish missions.

**Project Apollo**   The NASA program that concluded the U.S. effort to achieve the goal of landing a man on the moon and returning him safely to Earth.

**Project Gemini**   A manned spaceflight program conducted by NASA in 1966–67 to develop the necessary techniques of orbiting, docking two spacecraft in orbit, and extravehicular activity.

**Project Mercury**   The first of three successive NASA programs (Gemini and Apollo were the other two) whose aim was to put a man on the moon. Project Mercury put the first U.S. astronaut in space.

**Space station**   A permanently manned facility that NASA hopes to construct in Earth orbit during the 1990s.

**Wind tunnel**   A tunnellike passage through which air is blown at a certain velocity to determine the effects of wind pressure on an object placed in the passage.

# SELECTED REFERENCES

Anderson, Frank. *Orders of Magnitude: A History of NACA and NASA, 1915–1980.* Washington, DC: U.S. Government Printing Office, 1981.

Anderton, David A. *Sixty Years of Aeronautical Research, 1917–1977.* Washington, DC: U.S. Government Printing Office, 1978.

Brooks, Courtney, James Grimwood, and Loyd Swenson Jr., *Chariots for Apollo: A History of Manned Lunar Spacecraft.* Washington, DC: U.S. Government Printing Office, 1979.

Clarke, Arthur C. *The Coming of the Space Age.* New York: Meredith Press, 1967.

Collins, Michael. *Carrying the Fire: An Astronaut's Journeys.* New York: Farrar, Straus & Giroux, 1974.

———. *Liftoff: The Story of America's Adventure in Space.* New York: Grove Press, 1988.

Cortright, Edgar M. *Apollo Expeditions to the Moon.* Washington, DC: U.S. Government Printing Office, 1975.

Hacker, Barton, and James Grimwood. *On the Shoulders of Titans: A History of Project Gemini.* Washington, DC: U.S. Government Printing Office, 1977.

Hirsch, Richard, and Joseph Trento. *The National Aeronautics and Space Administration.* New York: Praeger, 1973.

Jones, Bessie Zaban. *Lighthouse of the Skies: A History of the Smithsonian Astrophysical Observatory.* Washington, DC: Smithsonian Institution Press, 1965.

Logsdon, John H. *The Decision to Go to the Moon: Project Apollo and the National Interest.* Cambridge, MA: MIT Press, 1970.

McDougall, Walter A. *Heavens and the Earth: A Political History of the Space Age.* New York: Basic Books, 1985.

National Aeronautics and Space Agency. *The First Twenty-Five Years: 1958–1983.* Washington, DC: U.S. Government Printing Office, 1983.

Swenson, Loyd, Jr., James Grimwood, and Charles Alexander. *This New Ocean: A History of Project Mercury*. Washington, DC: U.S. Government Printing Office, 1966.

Trento, Joseph J. *Prescription for Disaster: From the Glory of Apollo to the Betrayal of the Shuttle*. New York: Crown, 1987.

Wells, Helen, Susan H. Whitely, and Carrie Karegeannes. *Origins of NASA Names*. Washington, DC: U.S. Government Printing Office, 1976.

*We Seven: By the Astronauts Themselves*. New York: Simon and Schuster, 1962.

# INDEX

Department of Defense (DOD), 39, 45
Doolittle, James H., 34
Dryden, Hugh Latimer, 43, 51
Dryden Test Flight Center, 23, 26

*Early Bird* (satellite), 87
*Echo 1* (satellite), 51, 86, 87
Edwards Air Force Base, 23, 26
Eisele, Donn F., 77
Eisenhower, Dwight D., 17, 19, 39, 42, 43, 45, 53, 54, 59
England, 31
*Enterprise* (Shuttle Orbiter), 111
Environmental Sciences Services Administration (ESSA), 88
*Explorer 1,* 45

Faget, Maxime, 52, 53, 54
Fletcher, James C., 109
France, 31
*Freedom 7,* 57, 59
*Friendship 7,* 66

Gagarin, Yury Alekseyevich, 56, 57
Galileo program, 99
*Gemini 6A,* 69
*Gemini 7,* 68
*Gemini 8,* 68
Gemini program, 65, 67–69, 71
Geostationary Operational Environmental Satellite (GOES), 88
Germany, 31
Gilruth, Robert, 54, 62, 64
Glenn, John H., Jr., 55, 66–67, 78
Glennan, T. Keith, 43, 45, 54, 61, 97
Goddard, Robert, 128

Goddard Space Flight Center, 26, 50
Gorbachev, Mikhail, 127
Greenbelt, Maryland, 26, 50
Grissom, Virgil I. "Gus," 55, 66, 72, 75, 77

Hampton, Virginia, 22, 34
Harvard College Observatory, 15
Helios, 99
High Energy Astronomy Observatories, 99
Houston, Texas, 28, 62
Hubble Space Telescope, 99
Hughes Aircraft Corporation, 87
Hunsaker, Jerome C., 35
Huntsville, Alabama, 15, 26, 28, 63

Infrared Astronomy Satellite, 99
International Geophysical Year (IGY), 16
International Radiation Investigation Satellite, 99
International Sun-Earth Explorer, 99
International Telecommunications Satellite Consortium (Intelsat), 87
International Ultraviolet Explorer, 99
Iowa, University of, 45
Italy, 31
ITOS 1 (Improved Tiros Operational Satellite), 88

Japan, 36
Jet Propulsion Laboratory (JPL), 26, 50, 74
John C. Stennis Space Center, 28
Johns Hopkins University, 35

Johnson, Lyndon B., 15, 42, 61, 76
Johnson Space Center (JSC), 28, 62
Jupiter, 28, 96, 99
Jupiter (launch vehicle), 51
Jupiter-C (rocket), 45, 52

Kennedy, John F., 59, 60–62, 69, 79, 87
Kennedy Space Center, 28, 75
Kerr-McGee Company, 61
Kerwin, Joseph, 104
Khrushchev, Nikita, 61
Killian, James, 42
Knox, Frank, 36
Komarov, Vladimir, 71
Kosygin, Aleksey, 106

Lancaster, California, 23
Landsats (satellites), 89
Langley, Samuel Pierpont, 34–35
Langley Memorial Aeronautical Laboratory, 34
Langley Research Center, 22, 23, 38, 62
Latin America, 16
Lewis, George, 36
Lewis Research Center (LRC), 26
*Liberty Bell 7*, 66
Lindbergh, Charles, 34
Lovell, James, 67, 78
Lunar Module (LM), 64, 76, 78, 79, 81
Lunar Orbiter program, 75

McAuliffe, Christa, 114
McDonnell Aircraft, 54, 61
McElroy, Neil, 15
McNamara, Robert S., 61
Man In Space Soonest (MISS), 51
Manned Space Center, 28

*Mariner* (spacecraft), 91–93
Mars, 28, 61, 91, 92
Marshall, George C., 26
Marshall Space Flight Center (MSFC), 26
Massachusetts Institute of Technology, 35
Medaris, John, 16
Michaud Assembly Facility, 26
Mountain View, California, 23, 36
Muroc Dry Lake, 26, 38

National Advisory Committee for Aeronautics (NACA), 22, 26, 33–43
National Aeronautics and Space Act (1958), 19, 42, 53
National Aeronautics and Space Administration (NASA)
administration of, 20–21
Aeronautics and Space Technology, 21–22
Astronauts
influence on spacecraft design, 55
popularity of, 55
requirements of, 54
selection, 54
budget of, 19, 20
Configuration Control Board, 73
creation of, 19
and decision to go to the moon, 61
decline during Johnson administration, 76
and manned space effort, 59–69
Office of the Administration, 20
primary goal, 19
replaces NACA, 42–43
Space Flight, 21

138

**Tom D. Crouch** is chairman of the Department of Aeronautics of the Smithsonian Institution's National Air and Space Museum. Dr. Crouch, who has also served as chairman of the Department of Social and Cultural History of the National Museum of American History, holds a Ph.D. in history from Ohio State University. He has written and edited a number of books on the history of flight technology, including *The Bishop's Boys: A Life of Wilbur and Orville Wright* (1989), *The Eagle Aloft: Two Centuries of the Balloon in America* (1984), *The Bleriot XI: Story of a Classic Airplane* (1982), *A Dream of Wings* (1981), *Apollo XI: Ten Years Since Tranquility Base* (1979), and *Charles A. Lindbergh: An American Life* (1978).

**Arthur M. Schlesinger, jr.,** served in the White House as special assistant to Presidents Kennedy and Johnson. He is the author of numerous acclaimed works in American history and has twice been awarded the Putlizer Prize. He taught history at Harvard College for many years and is currently Albert Schweitzer Professor of the Humanities at the City College of New York.